T0224047

Communications in Computer and Information Science 1019

Commenced Publication in 2007
Founding and Former Series Editors:
Phoebe Chen, Alfredo Cuzzocrea, Xiaoyong Du, Orhun Kara, Ting Liu,
Krishna M. Sivalingam, Dominik Ślęzak, Takashi Washio, Xiaokang Yang,
and Junsong Yuan

Editorial Board Members

More information about this series at http://www.springer.com/series/7899

Chetan Arora · Kaushik Mitra (Eds.)

Computer Vision Applications

Third Workshop, WCVA 2018
Held in Conjunction with ICVGIP 2018
Hyderabad, India, December 18, 2018
Revised Selected Papers

 Springer

Editors
Chetan Arora
Department of Computer Science
and Engineering
Indian Institute of Technology Delhi
New Delhi, India

Kaushik Mitra
Department of Electrical Engineering
Indian Institute of Technology Madras
Chennai, India

ISSN 1865-0929   ISSN 1865-0937   (electronic)
Communications in Computer and Information Science
ISBN 978-981-15-1386-2   ISBN 978-981-15-1387-9   (eBook)
https://doi.org/10.1007/978-981-15-1387-9

This Springer imprint is published by the registered company Springer Nature Singapore Pte Ltd.
The registered company address is: 152 Beach Road, #21-01/04 Gateway East, Singapore 189721, Singapore

Preface

It is our pleasure to present the proceedings of the Third Workshop on Computer Vision Applications (WCVA 2018). The workshop was colocated with the 11th Indian Conference on Computer Vision, Graphics and Image Processing (ICVGIP), at the International Institute of Information Technology (IIIT) Hyderabad. Keeping in line with the traditions established from the past workshops, WCVA 2018 provided an appropriate platform for academic researchers as well as industry and government research labs to present and discuss their ideas on novel techniques for various computer vision applications.

The proceedings consists of 10 papers. All papers had an oral presentation during the workshop meeting on December 18, 2018. We thank the general chairs Prof. Rama Chelappa and Prof. Santanu Chaudury for their mentorship and all the reviewers for their careful and timely review. There were 32 papers submitted which were distributed to a technical program committee comprised of 31 well-qualified reviewers. All the papers were reviewed by 3 or more reviewers.

The WCVA program also included an inaugural talk by Prof. Santanu Chaudhury, Director IIT Jodhpur, and a keynote talk by Prof. Saket Anand from IIIT Delhi. The keynote talk covered an important area and was titled "Computer vision for Wildlife: from Conservation to Conflict Management." Afterwards, there were oral presentations of all the selected papers. We hope that the workshop papers capture the most important works being carried out by the Indian as well as the international researchers in important application areas of computer vision.

Finally, we thank the organizers of ICVGIP 2018 for providing the local logistic arrangement and administrative assistance, which was essential for the execution of the workshop. We would also like to thank all the authors for submitting their original work and all the participants for their interest and support.

December 2018

Chetan Arora
Kaushik Mitra

Organization

General Chairs

Rama Chellappa	University of Maryland, College Park, USA
Santanu Chaudhury	IIT Jodhpur, India

Program Chairs

Chetan Arora	IIT Delhi, India
Kaushik Mitra	IIT Madras, India

Technical Program Committee

Saket Anand	IIIT Delhi, India
Abir Das	IIT Kharagpur, India
Aditya Nigam	IIT Mandi, India
Gaurav Sharma	NEC Labs, USA
Anubha Gupta	IIIT Delhi, India
Ravikiran Sarvadevabhatla	IIIT Hyderabad, India
Prithwijit Guha	IIT Guwahati, India
Shanmuganathan Raman	IIT Gandhinagar, India
Suresh Sundaram	IIT Guwahati, India
Manoj Sharma	Rice University, USA
Jun-Cheng Chen	University of Maryland, College Park, USA
Vijay Rengarajan	CMU, USA
Rushil Anirudh	Lawrence Livermore National Laboratory, USA
Kuldeep Kulkarni	CMU, USA
Arnab Bhattacharya	IIT Kanpur, India
Sunil Simon	IIT Kanpur, India
Amay Karkare	IIT Kanpur, India
Vinay Namboodri	IIT Kanpur, India
Sumantra Dutta Roy	IIT Delhi, India
Gaurav Harit	IIT Jodhpur, India
Ayesha Chaudhury	JNU, India
Vikram Goyal	IIIT Delhi, India
Vineeth Balasubramanian	IIT Hyderabad, India
Anup Namboodri	IIIT Hyderabad, India
C. V. Jawahar	IIIT Hyderabad, India
Vineet Gandhi	IIIT Hyderabad, India
Parag Chaudhury	IIT Bombay, India
Uma Mudenagudi	BVB CET, Hubli, India
Soma Biswas	IISc, Bangalore, India

Venkatesh Babu	IISc, Bangalore, India
Parag Singla	IIT Delhi, India
Mansi Sharma	IIT Madras, India

Contents

Depth Augmented Semantic Segmentation Networks for Automated Driving

Hazem Rashed[1], Senthil Yogamani[2(✉)], Ahmad El-Sallab[1], Arindam Das[3], and Mohamed El-Helw[4]

[1] CDV AI Research, Cairo, Egypt
[2] Valeo Vision Systems, Valeo, Cairo, Egypt
[3] Detection Vision Systems, Valeo, Tuam, Ireland
{hazem.rashed,senthil.yogamani,
ahmad.el-sallab,arindam.das}@valeo.com
[4] Nile University, Cairo, Egypt
melhelw@nu.edu.eg

Abstract. In this paper, we explore the augmentation of depth maps to improve the performance of semantic segmentation motivated by the geometric structure in automotive scenes. Typically depth is already computed in an automotive system to localize objects and path planning and thus can be leveraged for semantic segmentation. We construct two networks that serve as a baseline for comparison which are "RGB only" and "Depth only", and we investigate the impact of fusion of both cues using another two networks which are "RGBD concat", and "Two Stream RGB+D". We evaluate these networks on two automotive datasets namely Virtual KITTI using synthetic depth and Cityscapes using a standard stereo depth estimation algorithm. Additionally, we evaluate our approach using monoDepth unsupervised estimator [10]. Two-stream architecture achieves the best results with an improvement of 5.7% IoU in Virtual KITTI and 1% IoU in Cityscapes. There is a large improvement for certain classes like trucks, building, van and cars which have an increase of 29%, 11%, 9% and 8% respectively in Virtual KITTI. Surprisingly, CNN model is able to produce good semantic segmentation from depth images only. The proposed network runs at 4 fps on TitanX GPU, Maxwell architecture.

Keywords: Semantic segmentation · Visual perception · Automated driving

1 Introduction

Recently, semantic segmentation has gained a huge attention in the field of computer vision. One of the main applications is autonomous driving where the car is able to understand the environment by providing a class for each pixel

© Springer Nature Singapore Pte Ltd. 2019
C. Arora and K. Mitra (Eds.): WCVA 2018, CCIS 1019, pp. 1–13, 2019.
https://doi.org/10.1007/978-981-15-1387-9_1

in the scene and consequently has the ability to react accordingly [14]. In this work, we investigate the usage of geometric cues to improve accuracy of semantic segmentation.

Most of semantic segmentation algorithms mainly rely on appearance cues and do not exploit geometry related information. In this paper, we investigate usage of depth as a geometric cue for semantic segmentation task in autonomous driving application where there is a strong geometric structure. The road surface is typically flat and all the objects stand vertical on it. This is exploited explicitly in the formulation of a commonly used depth representation namely Stixels [6]. The contributions of this work include:

1. Detailed study of the impact of depth for segmentation in automated driving.
2. Systematic study of fusing RGB and Depth on semantic segmentation using four CNN networks.
3. Experimentation on two automotive datasets namely Virtual KITTI and Cityscapes.

The rest of the paper is organized as follows: Sect. 2 reviews the related work in segmentation, depth computation and role of depth in semantic segmentation. Section 3 illustrates the details of our four architectures to systematically study the effect of fusing depth with appearance for semantic segmentation. Section 4 discusses the experimental results in Virtual KITTI and Cityscapes. Finally, Sect. 5 provides concluding remarks.

2   Related Work

2.1   Semantic Segmentation

Siam et al. [25] presented a detailed survey on automated driving particularly for semantic segmentation. The advancement of semantic segmentation until the present can be categorically discussed in three phases. It started with patch-wise training as reported in [8] for classification. [8] proposed multi-scale pyramid processing through 3-stage network followed by a classical segmentation approach as post processing. Grangier et al. [11] proposed a pixel level classification approach using deep network to avoid post processing but it could not remove patch-wise training.

Next level of progress was pixel-wise classification through end-to-end learning as reported in [1,18,22]. Fully convolutional network (FCN) [18] was the first deep learning based technique that did not use patch-wise training, rather it directly learned from the heatmaps. Series of upsampling layers were used to obtain the dense predictions. Later deconvolution layer was proposed in Segnet [1] in place of unpooling layer. Introduction of skip connection from encoder to decoder was another contribution to this work for output reconstruction.

Recently feature extraction from multi-scale input has been heavily explored and can be found in [4,8,22–24,31]. Though [8] used feature maps from encoder using skip connections to merge heatmaps from different resolution but space reduction in encoder side hurt the final prediction. U-net [24] pools encoded feature maps from initial layers that are concatenated with the decoded feature maps and upsampled for the next layers. To avoid loss of resolution, broadening the receptive field by applying dilated convolution has shown better results.

2.2 Depth in Automated Driving Systems

Depth estimation is very critical for automated driving. Having image semantics without localization is seldom useful. In a typical automated driving pipeline, depth is already computed and can be leveraged for semantic segmentation. In this sub-section, we summarize the different mechanisms by how depth can be estimated.

Classical Geometric Approach. Dense depth is computed to understand the spatial geometry of the scene. Stereo cameras have been commonly used in front camera automated driving systems. Disparity estimation methods using classical geometric matching algorithms are quite mature. Alternatively, Structure From Motion (SFM) approaches can be used for monocular cameras. But they suffer from issues like handling moving objects, focus on expansion, etc. Accurate Depth could be useful for semantic segmentation and could be passed on as an extra channel. However, SFM estimates are quite noisy and also the algorithm variations over time could affect the training of the network. But in [2] some cues from the noisy point-cloud was inferred to act as features for segmentation. The cues proposed were: height above the camera, distance to the camera path, projected surface orientation, feature track density, and residual reconstruction error. The work in [16] proposed a way of jointly estimating the semantic segmentation and structure from motion in a conditional random field formulation.

CNN Based Depth Estimation. In recent years, several CNN-based monocular depth estimation approaches are trained in a supervised way which requires a single input image with no assumptions about the scene geometry or types of objects which are present. For autonomous driving application, unsupervised methods are very beneficial due to the lack of reliable annotated datasets that have depth maps provided for outdoor driving scenes. Unsupervised depth estimation is an open point of research. [32] used temporal information of video sequence to capture depth while [11] referred to as "monoDepth" used left-right consistency for stereo images to train the network while the depth is estimated from monocular images in inference. We exploit this approach to generate depth maps for both Virtual KITTI, and Cityscapes datasets in our experiments.

LIDAR Sensors. LIDAR sensors provide depth estimation with better accuracy and range compared to camera based estimation algorithms. However, their measurements are sparse in the image lattice as illustrated in Fig. 1. This leads to problems in learning a dense convolutional neural networks features directly and requires handling of sparsity [28]. But they can be fused with camera based dense depth. The method in [21] fused a sparse LIDAR for semantic segmentation using elastic fusion [30]. Generally, this is a good research problem to be pursued as LIDAR is becoming a standard sensor in automated driving systems.

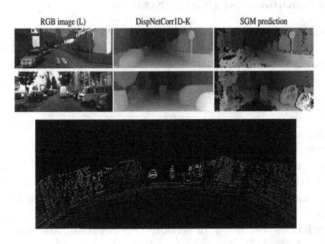

Fig. 1. Visualization of depth estimation (top) in automated driving scenes, adapted from [20]. It illustrates the output of a commonly used depth estimation algorithm called SGM we use in this paper and CNN based depth estimation which is closer to ground truth. Velodyne LIDAR depth re-projected on to a wide-angle image frame (bottom) to illustrate the level of sparsity.

2.3   Usage of Depth in Semantic Segmentation

FuseNet [12] is quite close to the work in this paper. They show that concatenating RGB and Depth slightly degrades mean IoU while the two-stream approach improves mena IoU by 3.65% in SUN RGB-D dataset. Ma et al. [19] combine depth and RGB for multi-view semantic segmentation where depth was leveraged to re-warp different views. Lin et al. [17] uses FCN based cascaded feature network with branch predictors and show an improvement of 2% in IoU compared to RGB baseline in NYU dataset. A detailed empirical study on role of depth for semantic segmentation and object detection was done in [3] and they show 2% improvement in IoU in VOC2012 dataset. Weiyue et al. [29] incorporate depth aware architecture design and obtain a larger improvement of 10% IoU in NYU dataset.

Apart from color, depth is another dimension and its influence for semantic segmentation task is relatively less explored. Above mentioned works that use RGB-D cameras are mainly focused for indoor scenes. On the other hand, different road conditions, diverse lighting states and presence of dense shadow make the automotive scenes very challenging for semantic segmentation however, better geometric structure for the scene is one thing to be exploited. From the extensive literature study, it appears that there is no systematic study done on the influence of depth for automotive scenes and this motivated our work.

3   Semantic Segmentation Models

In this section, the four architectures used in this paper are illustrated. (Figure 2(c)) shows RGBD network which is based on concatenation of RGB image and Depth map as a four layer input. (Figure 2(d)) shows the two stream RGB+D network. RGB-only and Depth-only are shown in (Fig. 2 (a), (b)), and they are used as baselines for comparison.

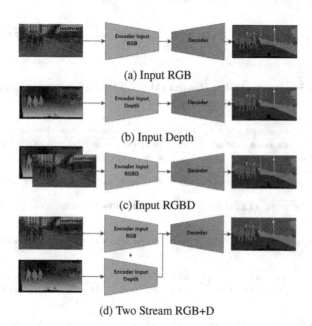

(a) Input RGB

(b) Input Depth

(c) Input RGBD

(d) Two Stream RGB+D

Fig. 2. Four types of architectures constructed and tested in the paper. (a) and (b) are baselines using RGB and Depth only. (c) and (d) are depth augmented semantic segmentation architectures.

Table 1. Quantitative analysis of our four networks on Virtual KITTI dataset.

Network type	IoU	Precision	Recall	F-Score
RGB	66.47	78.23	75.6	73.7
D	55	75.72	70.31	69.39
RGBD (concat)	66.76	77.6	72.4	72.3
RGB + D (2-stream add)	68.6	82.4	77.73	76.73
RGB + D (2-stream concat)	**72.13**	**82.5**	**79.96**	**79.59**

Table 2. Semantic Segmentation Results (Mean IoU) on Virtual KITTI dataset (GT - Ground Truth, mD - monoDepth)

Type	Mean	Truck	Car	Van	Road	Sky	Vegetation	Building	Guardrail	TrafficSign	TrafficLight	Pole
RGB	66.47	33.66	85.69	29.04	95.91	93.91	80.92	68.15	81.82	66.01	65.07	40.91
D (GT)	55	**67.68**	58.03	56.3	73.81	94.38	53.64	43.95	14.61	53.97	56.51	**42.67**
RGBD (GT)	66.76	65.34	91.74	**56.93**	95.46	**94.41**	78.17	54.91	73.42	60.21	46.09	30.46
RGB+D (GT-add)	68.6	43.38	91.59	29.19	96.01	94.32	85.17	77.6	80.13	69.54	**72.73**	32.09
RGB+D (GT-concat)	**72.13**	62.84	**93.32**	38.42	**96.33**	94.2	**90.46**	**79.04**	**90.85**	**72.22**	67.83	34.4
D (mD)	46.1	36.05	75.46	33.2	77.3	87.3	39.3	32.3	6.8	42.14	45.9	15.9
RGB+D (mD-add)	67.05	42.9	86.9	43.5	96.2	94.1	88.07	65.94	85.4	65.7	51.25	30.13
RGB+D (mD-concat)	68.92	40.57	86.1	50.3	95.95	93.82	81.63	70.43	86.3	68.66	67.58	35.94

Table 3. Semantic Segmentation Results (Mean IoU) on Cityscapes dataset (SGM - Semi Global Matching, mD - monoDepth)

Type	Mean	Bicycle	Person	Rider	Motorcycle	Bus	Car	Fence	Building	Road	Sidewalk	Sky	TrafficSign
RGB	62.47	63.52	**67.93**	40.49	29.96	62.13	89.16	44.53	87.86	96.22	74.98	89.79	59.88
D (SGM)	47.8	39.84	54.99	29.04	11.29	48.1	82.36	34.32	78.42	95.14	67.78	81.18	27.96
RGBD (SGM)	55.5	56.68	60.27	34.64	21.18	58	86.94	36.47	84.7	94.84	70.39	84.64	45.48
RGB+D (SGM-add)	**63.48**	**66.46**	67.85	**42.31**	41.37	63.1	89.77	**46.28**	**88.1**	96.38	75.66	90.23	**60.78**
RGB+D (SGM-concat)	63.13	65.32	67.79	39.14	37.27	**69.71**	**90.06**	42.75	87.44	**96.6**	**76.35**	**91.06**	59.44
D (mD)	40.89	36.63	44.6	18.5	7.3	37.5	77.78	16.16	77.01	92.83	54.87	89.33	24.67
RGB+D (mD-add)	61.39	66.23	67.33	39.9	44.01	55.7	89.1	40.2	87.34	96.47	75.7	88.7	57.7
RGB+D (mD-concat)	63.03	65.85	67.44	41.33	**46.24**	66.5	89.7	33.6	87.25	96.01	73.5	90.3	59.8

3.1 One-Stream Networks

This network is based on FCN8s [18] architecture and it's used in our RGB-only and Depth-only experiments. The fully connected layers of the VGG16 are changed to a fully convolutional network where the first 15 convolutional layers are used for feature extraction. The output segmentation decoder follows the FCN architecture where 1×1 convolutional layer is used followed by three transposed convolution layers for up-sampling. Introduction of skip connections within encoder was not tried as residual learning is not much effective for smaller networks as shown in [7]. Skip connections from encoder to decoder are exploited to extract high resolution features from the lower layers which are added to the upsampled feature maps. The loss function used for semantic segmentation is illustrated below.

$$L = -\frac{1}{|I|} \sum_{i \in I} \sum_{c \in C_{Dataset}} p_i(c) \log q_i(c) \tag{1}$$

where q denotes predictions and p denotes ground-truth. $C_{Dataset}$ is the set of classes for the used dataset.

3.2 RGBD Network

Four channels which are the original RGB image layers concatenated with the depth map are used as an input to the network, where depth layer is normalized from 0 to 255 to have the same value range as the RGB. The VGG pretrained weights are utilized, however the first layer is changed so that it accepts an input of four channels, where the corresponding weights are initialized randomly. Depth map Ground Truth is used in the case of Virtual KITTI to eliminate the errors due to depth estimation algorithms. For Cityscapes, disparity maps computed using SGM algorithm [13] are exploited where, it is a commonly used depth estimation algorithm in automated driving.

3.3 Two Stream (RGB+D) Network

Inspired from [15, 26, 27], a two-stream network using two VGG6 encoders is used, where each encoder processes a different input. One for the RGB input, and the other for the depth map. Fusion between feature maps from both encoders is done using two approaches. The first one is the usage of summation junction (RGB+D Add), while the other is concatenation instead of summation (RGB+D concat). By concatenation, depth dimension of the feature vector is doubled, however we aim to give the network more flexibility to learn more complex fusion approach to improve result. Afterwards, The same decoder used in the one-stream network is used for upsampling

4   Experiments

In this section, we present the datasets used, experimental setup and results.

4.1   Datasets

We choose two datasets namely Virtual KITTI [9] and Cityscapes [5] since they contain outdoor road scenes and this is consistent with our application as it is focused on automated driving. Additionally, Virtual KITTI provides perfect Ground Truth Depth annotation which helps us to evaluate our algorithm excluding the depth calculation errors. Typically, depth is calculated in automotive applications using stereo images or structure from motion. We utilize

Cityscapes depth maps which are based on SGM algorithm using stereo images. Virtual KITTI is a synthetic dataset that consists of 21,260 frames containing road scenes in an urban environments in different weather conditions. We exploit both depth and semantic segmentation annotations. Cityscapes is a well known dataset containing real images of road scenes. It consists of 20000 images having coarse semantic segmentation annotation and 5000 having fine annotation. We only use the fine annotations in our experiments and we intentionally use noisy SGM depth to understand the effects of relatively noise depth estimations, and we provide evaluation using IoU metric on the validation set that contains 500 frames.

4.2   Experimental Setup

We have used Virtual KITTI and Cityscapes dataset where the dimension of each image is 375×1242 and 1024×2048 (later down-scaled to 512×1024 during training) respectively. For all experiments, we transferred the encoder weights of VGG pre-trained model on ImageNet for the segmentation task. Transfer learning helped us to get better initialization of the encoder at the beginning of the joint encoder-decoder training for semantic segmentation. Dropout with probability 0.5 is used in our model particularly for 1×1 convolutional layers. Very popular Adam optimizer is used with an initial learning rate of $1e^{-5}$ along with L2 regularization in the loss function and a factor of $5e^{-4}$ to avoid over-fitting. To evaluate the efficacy of our proposal, widely used Intersection over Union (IoU) is measured for both datasets, also precision, recall and F-score are used for Virtual KITTI dataset.

4.3   Experimental Results

We provide qualitative evaluation on both datasets separately using IoU metric as shown in Tables 1, 2 and 3. Video links of the four architectures results are also provided for both datasets. In addition to depth annotations provided with the datasets, we generated depth maps using unsupervised approach [10] for both datasets and compared results against Ground Truth in Virtual KITTI, and noisy SGM in Cityscapes.

   Table 1 illustrates that depth augmentation consistently improves results in all four metrics reported. An improvement of 5.7% in IoU, 3.8% in Precision, 4.36% in Recall and 5.9% in F-score is shown. Class-wise evaluation is listed in Table 2. Although the overall improvement is incremental, there is a large improvement for certain classes, for example, trucks, van, Building and Traffic Lights show an improvement of 32%, 28%, 9% and 8% respectively. Cityscapes results are reported in Table 3, and it shows a relatively moderate improvement of 1% in IoU. However, results show that even noisy depth maps with invalid values due to depth estimation errors can improve semantic segmentation.

   In summary, the network that concatenates depth with RGB feature maps shows better results than others as observed on VKITTI. As per the results on CityScapes, the impact of feature map concatenation and addition with RGB is fairly close, however results from monoDepth [10] estimator using concatenated

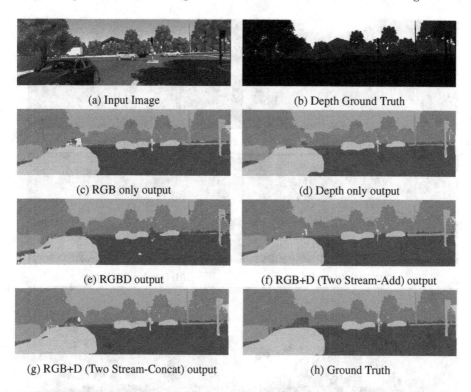

(a) Input Image   (b) Depth Ground Truth

(c) RGB only output   (d) Depth only output

(e) RGBD output   (f) RGB+D (Two Stream-Add) output

(g) RGB+D (Two Stream-Concat) output   (h) Ground Truth

Fig. 3. Qualitative comparison of semantic segmentation outputs from four architectures on VKITTI dataset using GT

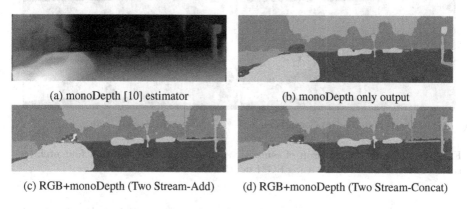

(a) monoDepth [10] estimator   (b) monoDepth only output

(c) RGB+monoDepth (Two Stream-Add)   (d) RGB+monoDepth (Two Stream-Concat)

Fig. 4. Qualitative comparison of semantic segmentation outputs using monoDepth [10] estimator

feature maps happened to outperform added feature maps. Qualitative results of all four proposals are demonstrated in on Virtual KITTI (Figs. 3 and 4) and Cityscapes (Figs. 5 and 6).

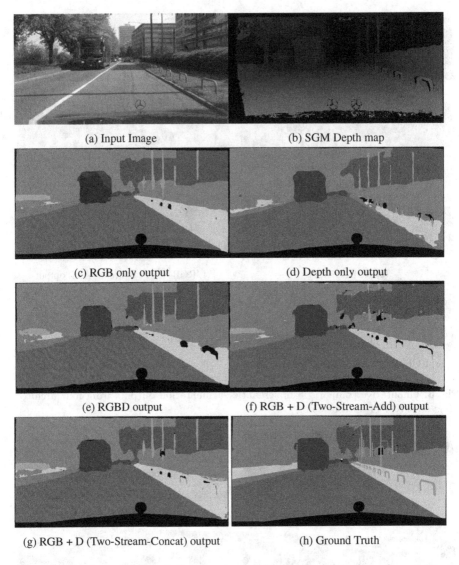

(a) Input Image   (b) SGM Depth map

(c) RGB only output   (d) Depth only output

(e) RGBD output   (f) RGB + D (Two-Stream-Add) output

(g) RGB + D (Two-Stream-Concat) output   (h) Ground Truth

Fig. 5. Qualitative comparison of semantic segmentation outputs from four architectures on Cityscapes dataset using SGM depth estimator

Test results of both datasets are shared publicly on YouTube in[1] and[2]. Depth-only network is reported to study the performance depth cue alone can do to semantic segmentation. Surprisingly, depth provides good results especially for road, vegetation, vehicle, and pedestrians. This is also consistent with the results obtained by [12] when depth only is tested for indoor scenes. We noticed that

---

[1] https://youtu.be/cfLUvE9knBU.
[2] https://youtu.be/vsYaVbcILbw.

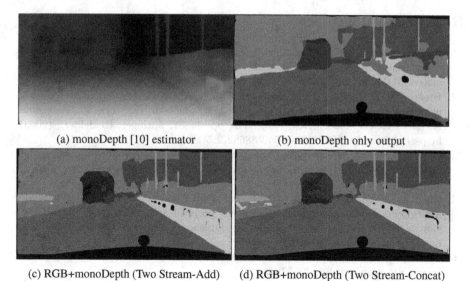

(a) monoDepth [10] estimator   (b) monoDepth only output

(c) RGB+monoDepth (Two Stream-Add)   (d) RGB+monoDepth (Two Stream-Concat)

Fig. 6. Qualitative comparison of semantic segmentation output from monoDepth estimator

there is degradation of accuracy relative to RGB baseline whenever there is noisy depth. Hence next step would be to make more systematic evaluation of the depth that is loosely coupled within the network. It is observed that the joint network has outperformed depth only network with a negligible margin, perhaps the network does not really know how to learn these two completely different cues and thus these two modalities are not logically fused in the network. Our future plan includes construction of multi-modal architectures to achieve better amalgamation of heterogeneous cues.

5   Conclusion

In this paper, we focused on the impact of a relatively unexplored cue that is depth for semantic segmentation task. We designed four different segmentation networks that receive input as RGB only, depth only, RGBD concatenated and two-stream RGB and depth. Our experimental results of four models on two automotive datasets namely Virtual KITTI and Cityscapes demonstrate a reasonable improvement in overall accuracy and good improvement for a few specific classes for the network that uses simple depth augmentation. We believe the present study furnishes adequate evidence on the impact of the depth for accurate semantic segmentation. In future work, we build a better depth aware more robust model to fully utilize its complementary nature.

References

1. Badrinarayanan, V., Kendall, A., Cipolla, R.: SegNet: a deep convolutional encoder-decoder architecture for image segmentation. arXiv preprint arXiv:1511.00561 (2015)
2. Brostow, G.J., Shotton, J., Fauqueur, J., Cipolla, R.: Segmentation and recognition using structure from motion point clouds. In: Forsyth, D., Torr, P., Zisserman, A. (eds.) ECCV 2008. LNCS, vol. 5302, pp. 44–57. Springer, Heidelberg (2008). https://doi.org/10.1007/978-3-540-88682-2_5
3. Cao, Y., Shen, C., Shen, H.T.: Exploiting depth from single monocular images for object detection and semantic segmentation. IEEE Trans. Image Process. **26**(2), 836–846 (2017)
4. Chen, L.C., Yang, Y., Wang, J., Xu, W., Yuille, A.L.: Attention to scale: Scale-aware semantic image segmentation. arXiv preprint arXiv:1511.03339 (2015)
5. Cordts, M., et al.: The cityscapes dataset for semantic urban scene understanding. arXiv preprint arXiv:1604.01685 (2016)
6. Cordts, M., et al.: The stixel world: a medium-level representation of traffic scenes. Image Vis. Comput. **68**, 40–52 (2017)
7. Das, A., Yogamani, S.: Evaluation of residual learning in lightweight deep networks for object classification. In: Proceedings of the Irish Machine Vision and Image Processing Conference, pp. 205–208 (2018)
8. Farabet, C., Couprie, C., Najman, L., LeCun, Y.: Learning hierarchical features for scene labeling. IEEE Trans. Pattern Anal. Mach. Intell. **35**(8), 1915–1929 (2013)
9. Gaidon, A., Wang, Q., Cabon, Y., Vig, E.: Virtual worlds as proxy for multi-object tracking analysis. In: CVPR (2016)
10. Godard, C., Mac Aodha, O., Brostow, G.J.: Unsupervised monocular depth estimation with left-right consistency. In: CVPR, vol. 2, p. 7 (2017)
11. Grangier, D., Bottou, L., Collobert, R.: Deep convolutional networks for scene parsing. In: ICML 2009 Deep Learning Workshop, vol. 3. Citeseer (2009)
12. Hazirbas, C., Ma, L., Domokos, C., Cremers, D.: FuseNet: incorporating depth into semantic segmentation via fusion-based CNN architecture. In: Lai, S.-H., Lepetit, V., Nishino, K., Sato, Y. (eds.) ACCV 2016. LNCS, vol. 10111, pp. 213–228. Springer, Cham (2017). https://doi.org/10.1007/978-3-319-54181-5_14
13. Hirschmuller, H.: Accurate and efficient stereo processing by semi-global matching and mutual information. In: IEEE Computer Society Conference on Computer Vision and Pattern Recognition CVPR 2005, vol. 2, pp. 807–814. IEEE (2005)
14. Horgan, J., Hughes, C., McDonald, J., Yogamani, S.: Vision-based driver assistance systems: survey, taxonomy and advances. In: 2015 IEEE 18th International Conference on. Intelligent Transportation Systems (ITSC), pp. 2032–2039. IEEE (2015)
15. Jain, S.D., Xiong, B., Grauman, K.: Fusionseg: learning to combine motion and appearance for fully automatic segmention of generic objects in videos. arXiv preprint arXiv:1701.05384 (2017)
16. Kundu, A., Li, Y., Dellaert, F., Li, F., Rehg, J.M.: Joint semantic segmentation and 3D reconstruction from monocular video. In: Fleet, D., Pajdla, T., Schiele, B., Tuytelaars, T. (eds.) ECCV 2014. LNCS, vol. 8694, pp. 703–718. Springer, Cham (2014). https://doi.org/10.1007/978-3-319-10599-4_45
17. Lin, D., Chen, G., Cohen-Or, D., Heng, P.A., Huang, H.: Cascaded feature network for semantic segmentation of RGB-D images. In: 2017 IEEE International Conference on Computer Vision (ICCV), pp. 1320–1328. IEEE (2017)

18. Long, J., Shelhamer, E., Darrell, T.: Fully convolutional networks for semantic segmentation. In: Proceedings of the IEEE Conference on Computer Vision and Pattern Recognition, pp. 3431–3440 (2015)
19. Ma, L., Stückler, J., Kerl, C., Cremers, D.: Multi-view deep learning for consistent semantic mapping with RGB-D cameras. arXiv preprint arXiv:1703.08866 (2017)
20. Mayer, N., et al.: A large dataset to train convolutional networks for disparity, optical flow, and scene flow estimation. In: Proceedings of the IEEE Conference on Computer Vision and Pattern Recognition, pp. 4040–4048 (2016)
21. McCormac, J., Handa, A., Davison, A., Leutenegger, S.: Semanticfusion: dense 3D semantic mapping with convolutional neural networks. arXiv preprint arXiv:1609.05130 (2016)
22. Noh, H., Hong, S., Han, B.: Learning deconvolution network for semantic segmentation. In: Proceedings of the IEEE International Conference on Computer Vision. pp. 1520–1528 (2015)
23. Qi, G.J.: Hierarchically gated deep networks for semantic segmentation. In: The IEEE Conference on Computer Vision and Pattern Recognition (CVPR), June 2016
24. Ronneberger, O., Fischer, P., Brox, T.: U-Net: convolutional networks for biomedical image segmentation. In: Navab, N., Hornegger, J., Wells, W.M., Frangi, A.F. (eds.) MICCAI 2015. LNCS, vol. 9351, pp. 234–241. Springer, Cham (2015). https://doi.org/10.1007/978-3-319-24574-4_28
25. Siam, M., Elkerdawy, S., Jagersand, M., Yogamani, S.: Deep semantic segmentation for automated driving: taxonomy, roadmap and challenges. arXiv preprint arXiv:1707.02432 (2017)
26. Siam, M., Mahgoub, H., Zahran, M., Yogamani, S., Jagersand, M., El-Sallab, A.: MODNET: moving object detection network with motion and appearance for autonomous driving. arXiv preprint arXiv:1709.04821 (2017)
27. Simonyan, K., Zisserman, A.: Two-stream convolutional networks for action recognition in videos. In: Advances in Neural Information Processing Systems, pp. 568–576 (2014)
28. Uhrig, J., Schneider, N., Schneider, L., Franke, U., Brox, T., Geiger, A.: Sparsity invariant CNNs. arXiv preprint arXiv:1708.06500 (2017)
29. Wang, W., Neumann, U.: Depth-aware CNN for RGB-D segmentation. arXiv preprint arXiv:1803.06791 (2018)
30. Whelan, T., Leutenegger, S., Salas-Moreno, R.F., Glocker, B., Davison, A.J.: Elasticfusion: Dense slam without a pose graph. In: Robotics: Science and Systems, vol. 11 (2015)
31. Yu, F., Koltun, V.: Multi-scale context aggregation by dilated convolutions. arXiv preprint arXiv:1511.07122 (2015)
32. Zhou, T., Brown, M., Snavely, N., Lowe, D.G.: Unsupervised learning of depth and ego-motion from video. In: Proceedings of the IEEE Conference on Computer Vision and Pattern Recognition, pp. 1851–1858 (2017)

Optic Disc Segmentation in Fundus Images Using Anatomical Atlases with Nonrigid Registration

Ambika Sharma[1]([✉]), Monika Aggarwal[1]([✉]), Sumantra Dutta Roy[1]([✉]),
Vivek Gupta[2]([✉]), Praveen Vashist[2]([✉]), and Talvir Sidhu[2]([✉])

[1] Indian Institute of Technology Delhi (IITD), New Delhi, India
ambikasharma1108@gmail.com, moaggarwal@yahoo.co.in, sumantra@ee.iitd.ac.in
[2] All Indian Institute of Medical Sciences (AIIMS) Delhi, New Delhi, India
vgupta@aiims.edu, applepv@gmail.com, talviraiims@gmail.com

Abstract. According to a WHO report, approximately 253 million people live with vision impairment, 36 million of which are blind and 217 million have moderate to severe vision impairment. In a recent estimate, the major causes of blindness are Cataract, Uncorrected refractive index, and Glaucoma. Thus in medical diagnosis, the retinal image analysis is a very vital task for the early detection of eye diseases such as Glaucoma, diabetic retinopathy (DR), Age-macular Degeneration (AMD) etc. Most of these eye diseases, if not diagnosed at an early stage might lead to permanent loss of vision.

A critical element in the computer-aided diagnosis of Digital Fundus images is the automatic detection of the optic disc region. Especially for the Glaucoma case, where cup to disc diameter ratio (CDR) is the most important indicator for detection. In this paper, we present a nonrigid registration based robust optic disc segmentation method using image retrieval based optic disc model maps that detect optic disc boundaries and surpasses the state-of-the-art performances. The proposed method consists of three main stages: (1) a content-based image retrieval from the model maps of OD using Bhattacharyya shape similarity measure, (2) constructing the test image specific anatomical model using the SIFT-flow technique for deformable registration of training masks to the test image OD mask, and (3) extracting the optic disc boundaries using a thresholding approach and smoothen the image by applying morphological operations along with the final ellipse fitting. The proposed work has used three datasets RIM, DRIONS and DRISHTI with 835 images in total. Our average accuracy values for 685 test images is 95.8%. The other performance parameter values are Specificity is 95.54%, Sensitivity is 96.13%, Overlap is 86.46% and Dice metric is 0.924 respectively, which clearly demonstrates the robustness of our optic disc segmentation approach.

Keywords: Computer vision · Image registration · Retinal image · Optic disc · Computer-aided detection (CAD) · Morphology · Optic disc (OD)

© Springer Nature Singapore Pte Ltd. 2019
C. Arora and K. Mitra (Eds.): WCVA 2018, CCIS 1019, pp. 14–27, 2019.
https://doi.org/10.1007/978-981-15-1387-9_2

1   Introduction

According to current eye disease statistics more than 42 million people are currently blind in the world, 80% of which could have been prevented or cured by early detection [1,14]. India is the second most populous country in the world and shares 17.5% of the world's population. Thus, in case of any health problem lead to a rapid increase in global morbidity rate. Currently, India has more than 15 million blind people, which is expected to increase to 16 million by 2020 [3,14]. Some of the most prevailing eye diseases are Cataract, Glaucoma, Diabetic retinopathy and Age-related macular degeneration. Glaucoma is the second leading cause of blindness in the world. It is also called silent-thief as the progression of disease is gradual and one might not able to diagnose it at early stage. In developing countries most of the population lives in remote and rural areas, therefore, it is not possible to reach out with due to a limited number of trained opticians and resources. As mentioned in [15], that eight minutes per eye is needed for complete segmentation of optic disc and cup. Thus, there is great need to have a cost-effective automatic computer-based diagnostic systems to enable even the people in remote and rural areas to get a medical diagnosis in time. Also, this CAD system will provide a preliminary evaluation of the patient's eye. Using Internet of things (IOT) based techniques we can further develop a system where the patients periodically test their eye while sitting at home using a handheld device and diagnostic system will evaluate the patient parameters which can further be uploaded into the virtual cloud and finally can be retrieved by the professionals at any time. This not only saves the doctor's time and effort but makes the procedure plat-form independent i.e. it will be able to work under different environment conditions where no instructions from a medical practitioner are necessary and able to interpret the results sensibly.

For the screening of most of the eye diseases, the detection and segmentation of optic disc is an important step. For e.g. in Glaucoma professionals look for the cup to disc diameter ratio (CDR) as the key parameter for the diagnosis. The optic nerve examination includes the analysis of a fundus (retinal) image, which is the photograph of the inner surface of the eye opposite to lens and includes different anatomical structures (features) like retina, optic disc, macula, fovea and blood vessels. In a healthy fundus image with good contrast and resolution, segmentation of the optic disc is a tractable problem, but the situation becomes difficult when the pathological condition occurs. In a diseased fundus image, the contrast is no longer uniform and segmenting the region of interest becomes a challenging task.

A number of methods have been proposed for the optic disc segmentation. Some of the traditional image processing techniques used template matching approach along with highly saturated intensity in red channel for disc segmentation [3,4]. Abdel-Ghafar et al. [18] proposed a simple segmentation technique based on the edge detector and the circular Hough transform (CHT). The method uses the green channel for processing as it has the highest contrast and morphological operations are used to remove the blood vessels. After applying the "Sobel operator" to the green channel, the image is thresholded and resulting points are given as input to the circular Hough transform. The algorithm claims that the largest circle was consistently found to be an optic disc with its center as the approximated OD center. A watershed-based OD segmentation approach is proposed by Welfer et al. [19]. Other methods have used circular Hough Transform [8] and region growing techniques [10] with a prior knowledge of seed point in the region of interest. Active shape model [10] based techniques such as changes method for active contours [10] have been very popular in the medical imaging, but they fail to extract the exact boundary of optic disc in case of low gradient between optic disc and background and when the PPA region is present around OD which has the same color characteristics as that of optic disc. Also, AC based methods often fail to control the contour formation process as they either terminate far outside or inside the OD boundary. Additional challenges include segmenting the low quality and blurred images, making allowances for anatomical variations in resolution, contrast and optic disc inhomogeneities. Figure 1 shows some examples of such variations like poor contrast and PPA.

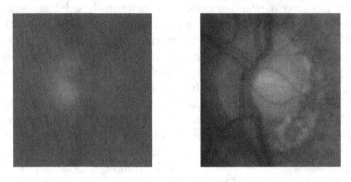

Fig. 1. Example of blurred and pathological optic disc (Color figure online)

In this paper for optic disc segmentation, we presented a robust automated optic disc segmentation system for retinal images. Our method mainly consists of three main stages as shown in the proposed architecture in Fig. 2. In the first subsection of the method we have build a anatomical model maps for optic disc with pre-segmented masks being marked by experts. The top 5 similar masks have been selected based on similarity coefficient between test image and model masks. The highly ranked masks retrieved by this method are usually a good fit for the test fundus image. In the second subsection, for the chosen masks the method first calculates the corresponding pixels between the test image and each of the model images which provides the transformation mapping for each of the pixel. Finally, it aligns the model masks using the transformation mapping. In the last subsection a thresholding has been applied to the combined segmented masks obtained by summing the outputs from each of the model map transformation. In order to smooth the optic disc boundary morphological closing has been used along with the ellipse fitting as optic disc is slightly vertically oval as per the literature. A detailed assessment of the approach compared to other state-of-the-art methods have been discussed in the later sections. In Fig. 2 the test image has been compared with the Atlas images using projection profile knowledge and the best five masks have been selected based on Bhattacharyya coefficient. Finally, SIFT features comparison is done between target image and given image in order to warp the model mask into the desired test optic disc mask. In the last step, the obtained probabilistic mask has been thresholded to get the segmented binary image. The proposed approach has huge advantage for the OD segmentation as it uses both prior knowledge of optic disc (unlike other model based approaches e.g. active contour) and require less amount of images in the atlas dataset (unlike machine learning models). Moreover as medical images are subjective in nature, this approach has helped us in handling this subjectivity.

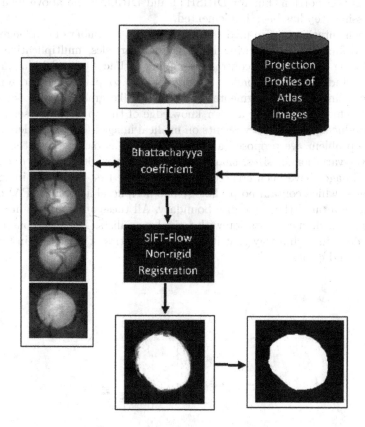

Fig. 2. Proposed architecture for optic disc segmentation

2   Proposed Method

2.1   CBIR Model from the Fundus Atlas for Inter-Image Matching

The Optic disc is a bright circular region in a fundus image. According to literature [3], disc is the brightest portion in the red channel of the retinal image. For
the extraction of disc boundary, we have used a pre-processing step of cropping
the Region of Interest from the complete retinal image and to do so we have
used the work of [5] on optic disc detection i.e. finding a point inside the optic
disc boundary, in order to have a initial seed point for cropping the region for
the optic disc. The algorithm uses the vessel convergence property at the optic
disc along with the disc characteristics (shape, size, and colour) to find a point
inside the disc boundaries. The RIM dataset contains the cropped region having ROI as OD portion, but for DRISHTI and DRIONS the above mentioned
pre-processing step has been implemented.

The segmentation task in medical imaging poses a number of challenges such
as light artifacts while capturing images, dust particles, multiplicative noise,
motion during imaging, low contrast, sampling artifacts caused by acquisition
equipment and finally anatomical variations due to pathological conditions.
Therefore, using classical segmentation methods like gradient, and thresholding based, which do not have a prior knowledge of the object to be segmented,
usually produce unsatisfactory results on medical images. Thus in order to solve
the above problem, we proposed a retinal image atlas dataset into the system
which gives variation in sizes, shapes, position with respect to the optic disc.
The atlas images have been created by selecting the best optic disc images i.e.
those images which contain no pathology like peripheral atrophy (PPA) or disc
haemorrhage around the optic disc boundary. All these atlas images have been
manually labelled, in consultation with retinal specialists. The atlas construction
has been done in such a way that it covers all varieties of optic disc in terms of
shape, size and color.

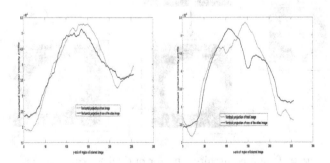

Fig. 3. Plot shows the horizontal (left) and vertical (right) projection profiles of test
and one of the atlas image

For the test image, we first identify a subset of images (i.e. five in our case)
from the model maps that are most similar to the test fundus ROI image, using

a content-based image retrieval (CBIR) inspired approach, and use this subset of training images including their corresponding OD masks to develop a test image specific OD model. The content based image retrieval has been done by calculating the intensity projection profiles along vertical and horizontal directions. For the atlas images the horizontal and vertical projection profiles are pre-computed in order to speed up the CBIR search process. For the similarity measurement between two distributions of test image and the atlas images Bhattacharyya coefficient has been used, which is as follows:

$$BC(I_1, I_2) = \alpha \sum_{x=1}^{n} \sqrt{p_1(x)p_2(x)} + (1 - \alpha) \sum_{y=1}^{m} \sqrt{q_1(y)q_2(y)} \qquad (1)$$

where $p_1(x)$ and $p_2(x)$ are the horizontal projections, $q_1(y)$ and $q_2(y)$ are the vertical projections of images I_1 and I_2 images respectively, x and y are the histogram bins of the projection profiles and n and m are the number of bins of the projection profiles and profile histograms and $\alpha = \frac{n}{n+m}$. The value of α varies for 0 to 1, but for the experimental work we have $n = m$ and thus results into $\alpha = 0.5$.

Figure 3 shows the horizontal and vertical profile histograms of an example image. Left image shows the horizontal projection profile between the test image and one of the best matching atlas image and right image shows the vertical projection profile respectively. The other distance metrics used in literature [28] for similarity measurement are Euclidean, Manhattan etc. but the chosen coefficient give the best possible performance under given conditions. We select a set of best fit training atlases from the anatomical database of segmented optic disc images to learn a test specific OD model. The registration performance for our method is significantly improved when a personalized OD model is designed by comparing the test ROI with the pre-segmented optic disc images in the database using a fast similarity measure based on Bhattacharyya coefficient.

In our proposed work, the similarity index has been calculated for both red and green channels, as red channel represents the saturated optic disc region for healthy images and thus works for most of the cases, but for abnormal conditions, the green channel performs better. Thus, the best channel has been selected based on the performance parameter value being calculated.

2.2 SIFT-Flow Deformable Warping of OD Atlas

Image registration scheme calculates a transformation mapping from the source image to target image by matching corresponding pixels of the images. The local image feature descriptors such as Scale Invariant Feature Transform (SIFT) [25], Histogram of Gradient (HOG) [26], shape and curvature descriptors can be used to match the correspondence. In this work, we used the SIFT descriptor which is among the best performing local image feature descriptors.

In computer vision image alignment remains a difficult task and the goal becomes even more difficult in the object recognition scenario, where the goal is to align different instances of the same object category. Similar to well known

Fig. 4. SIFT descriptors for test image (1st column) and atlas image (2nd column) respectively.

computer vision image alignment technique of optical flow where an image is aligned to its temporally adjacent frame, we used SIFT flow [27], a method to align an image to its nearest neighbors in a database containing a variety of objects. The SIFT features allow robust matching across different scene/object appearances, whereas the discontinuity preserving spatial model allows matching of objects located at different parts of the scene. Experiments show that the proposed approach robustly aligns complex scene pairs containing significant spatial differences. Our work is focused on inter-image similarity with deformable warping for creating a test image specific OD atlas. We found that the SIFT-flow algorithm worked well for this task. The SIFT features of the ROI are calculated as follows. First, image gradient magnitude and orientation are computed at each pixel. The gradients are weighted by a Gaussian pyramid in a $K \times K$ region. Then the regions are subdivided into $k \times k$ quadrant. The histogram of gradient orientations is calculated for 8 bins for each of the quadrant. Finally the orientation histograms are concatenated to construct the SIFT descriptor for the center pixel in all $K \times K$ regions. In definition of SIFT descriptors K, and k are chosen to be 16 and 4 respectively [27], thus for each pixel we have a 128 dimensional feature vector. We have shown in Fig. 4 two such SIFT images (also called per-pixel sift descriptor) corresponding to test image (on left) and an atlas image (on right side).

Once the SIFT descriptors haven been calculated for the image, the registration algorithm computes the correspondence between the test image and the atlas image by matching the SIFT descriptors. The SIFT flow algorithm consists of matching densely sampled, pixel-wise SIFT features between two images, while preserving spatial discontinuities [27].

The algorithm applies the transformation mapping by shifting each pixel in the atlas OD masks according to the calculated shift distances being given by the flow vectors. The registration stage is repeated for each of the top chosen masks (5 in our case). The obtained OD mask for the test image is calculated by adding all the transformed masks from each of the atlas OD regions and each pixel in the image represents the confidence level of the specific pixel belonging to the optic disc region.

2.3 Thresholding with Mask Smoothing

The obtained probabilistic masks can be smoothed further in order to enhance the robustness of the method. A smoothing filter is then applied on the decision values to achieve a smoothed decision value. In our implementation, mean filter and Gaussian filter are tested and the mean filter is found to be a better choice for the case. The smoothed decision values are then used further to calculate the binary decisions for all pixels using a threshold. A threshold value of 0.7 has been calculated empirically for all set of test images.

In our experiments, we have assigned a +1 and 0 to the disc (object) and non-disc (background) samples. An image closing morphological operations has been applied for disc shaped structuring element in order to remove the spikes present at disc boundary. At last, for smooth and continuous boundary ellipse fitting is done to the segmented optic disc region. Mostly medical experts label the optic disc as smooth curve, and to get that smoothness ellipse has been fitted. In fact this fitting has not changed the performance to a large extend (a little improvement by 0.3% in accuracy has been observed after ellipse fitting).

3 Experimental Results

3.1 Digital Retinal Image Datasets

The proposed method is evaluated using three different retinal databases, these are DRISHTI-GS1 dataset of 101 images [13,16] provided by Medical Image Processing (MIP) group, IIIT Hyderabad, DRIONS dataset of 110 images and finally RIM (RIM-1 and RIM-2) dataset of 624 images. In DRISHTI-GS1 dataset all images were taken with the eyes dilated, centered on OD with a Field-of-View of 30-degrees and of dimension 2896 × 1944 pixels and PNG uncompressed image format. The optic disc has been marked by experts for all 101 images.

The DRIONS database consists of 110 colour digital retinal images. The images were acquired with a colour analogical fundus camera, approximately centred on the ONH and they were stored in slide format. In order to have the images in digital format, they were digitised using a HP-PhotoSmart-S20 high-resolution scanner, RGB format, resolution 600 × 400 and 8 bits/pixel. The optic disc annotations have been done by two medical experts using a software tool.

Finally the RIM-1 database contains 169 optic nerve head images and each image has 5 manual segmentation from ophthalmic experts. The RIM-2 database consists of 455 images with disc annotated by the experts.

Figure 5 shows a subset of test images with the expert labeling in black color along with segmented OD boundary in green color respectively. The proposed method gives pretty good performance for DRISHTI and DRIONS datasets. Also for the RIM database which contains most of the PPA, blurred and poor intensity images the method gives satisfactory performance as shown in Fig. 5 last row images. The key advantage of the proposed method is that it works for inter database images i.e. in RIM database most of the portion of ROI is covered by the OD region whereas in DRIONS and DRISHTI the OD takes a small portion of the complete ROI. So, in spite of the OD size variability for a fixed image dimension the proposed method is able to extract very good estimation of the true OD boundary.

For the designing of atlas retinal images we have selected a subset of images from each of the datasets. The anatomical atlas consists of 85 RIM images from 624 images, 35 DRISHTI-GS images from a set of 101 images and 31 images from the 110 DRIONS images respectively. Thus in total 150 retinal images have been selected for the atlas model and 724 images is used for testing purpose.

3.2 Evaluation Metrics

In the literature several algorithms have proposed different evaluation metrics for the segmentation purpose. The validation metrics True positive (TP), True negative (TN), False positive (FP), and False negative (FN) have been used for verifying the quality of segmented image. Here TP, FP, TN, FN represents the pixels correctly classified as foreground, falsely classified as foreground, correctly detected as background, and falsely detected as background respectively. All the metrics used have been calculated pixel-wise. In our work of comparing the performance of proposed method with the state-of-the-art we have used these above metrics to find the Accuracy, Specificity, Sensitivity, Region Overlap and Dice metric. Their mathematical expressions have been given below:

$$(ACC)Accuracy(A,B) = \frac{(TP+TN)}{(P+N)} * 100 \tag{2}$$

$$(SPE)Specificity(A,B) = \frac{TN}{(TN+FP)} * 100 \tag{3}$$

$$(SEN)Sensitivity(A,B) = \frac{TP}{(TP+FN)} * 100 \tag{4}$$

$$(DM)DiceMetric(A,B) = \frac{2*TP}{FP+2*TP+FN} \tag{5}$$

$$(OL)RegionOverlap = \frac{TP}{(TP+FN+FP)} * 100 \tag{6}$$

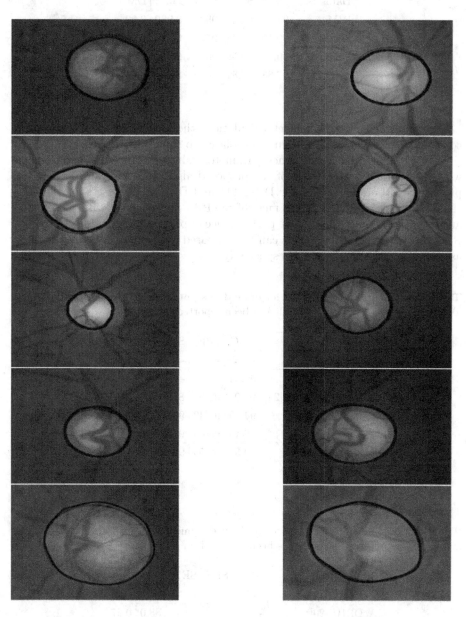

Fig. 5. Optic disc segmentation results. Here green and black colour represents the proposed and expert boundary respectively (Color figure online)

Table 1. Proposed algorithm performance parameters for different databases

Datasets	ACC	SPE	SEN	OL	DM
RIM	94.89	94.49	95.95	84.89	0.92
DRISHTI	99.15	99.43	97.38	93.44	0.96
DRIONS	99.30	99.50	96.55	90.64	0.95
AVG Perf.	97.88	97.81	96.56	89.66	0.95

For Bhattacharyya coefficient calculation, the value of $\alpha = 0.5$ has been considered as the image has been interpolated to square image.

Table 1 shows the performance parameter values for all the datasets along with the average performance of the proposed algorithm. We can see that the proposed method works well for DRISHTI and DRIONS datasets and also for the RIM dataset which contains most of the PPA images along with the blurred and intensity artifacts one's. The performance of proposed method for DRIONS, RIM and DRISHTI-GS datasets can be compared with state-of-the-art methods as shown in Tables 2, 3, and 4 respectively.

Table 2. Comparison of methods for optic disc segmentation for DRIONS database. The symbol "-" represents no result has been reported for the case

Methods	ACC	SPE	SEN	OL	DM
Walter et al. [22]	-	-	-	-	0.612
Morales et al. [23]	99.34	-	-	-	0.9084
CHT and Graph cut [24]	95.0	99.0	85.0	85.0	0.91
DRIU [20]	94.89	94.49	95.95	84.89	0.92
Zilly et al. [21]	99.15	99.43	97.38	93.44	0.96
Proposed method	99.3	99.5	96.45	90.64	0.95

Table 3. Comparison of methods for optic disc segmentation for RIM database. The symbol "-" represents no result has been reported for the case

Methods	ACC	SPE	SEN	OL	DM
Lu's [9]	91.0	-	-	-	-
DRIU [20]	-	-	-	88.0	0.97
Zilly et al. [21]	-	-	-	89.0	0.94
Proposed method	96.21	97.9	92.33	88.0	0.94

Table 4. Comparison of methods for optic disc segmentation for DRISHTI-GS database. The symbol "-" represents no result has been reported for the case

Methods	ACC	SPE	SEN	OL	DM
DRIU [20]	-	-	-	88.0	0.97
Zilly et al. [21]	-	-	-	-	-
A Sev. [11]	-	-	-	89.0	0.94
Joshi et al. [13]	-	-	-	-	0.96
Proposed method	99.15	99.42	97.28	93.44	0.96

4 Conclusion

We have presented a robust optic disc boundary detection method that is based on a test image specific atlas using the projection profile similarity selection and SIFT-flow nonrigid registration with refinement using filter smoothing and thresholding. We evaluated the algorithm on 712 test images with normal and pathological optic disc regions using three different databases. The experimental results showed an accuracy of 95.8% compared to expert segmentation gold standard. The other performance parameter values are Specificity is 95.54%, Sensitivity is 96.13%, Overlap is 86.46% and Dice metric is 0.924 respectively.

References

1. World Health Organization: Media centre: visual impairment and blindness, pp. 2–5 (2014). http://www.who.int/mediacentre/factsheets/fs282/en/
2. Hoover, A., Goldbaum, M.: Locating the optic nerve in a retinal image using the fuzzy convergence of the blood vessels. IEEE Trans. Med. Imaging **22**(8), 951–958 (2003)
3. Lowell, J., et al.: Optic nerve head segmentation. IEEE Trans. Med. Imaging **23**(2), 256–264 (2004)
4. Kumar, V., Sinha, N.: Automatic optic disc segmentation using maximum intensity variation. In: IEEE 2013 Tencon - Spring, TENCON Spring 2013 - Conference Proceedings, pp. 29–33 (2013)
5. Sharma, A., Agrawal, M., Lall, B.: Optic disc detection using vessel characteristics and disc features. In: 2017 Twenty-Third National Conference on Communications (NCC), Chennai, pp. 1–6 (2017). https://doi.org/10.1109/NCC.2017.8077135
6. Yin, F., et al.: Automated segmentation of optic disc and optic cup in fundus images for glaucoma diagnosis. In: Proceedings of the Symposium on Computer-Based Medical Systems (2012)
7. Wong, D.K., et al.: Level-set based automatic cup-to-disc ratio determination using retinal fundus images in ARGALI. In: Processings Conference on IEEE Engineering in Medicine and Biology Society, vol. 2008, no. 2, pp. 2266–2269 (2008)
8. Gopalakrishnan, A., Almazroa, A., Raahemifar, K., Lakshminarayanan, V.: Optic disc segmentation using circular hough transform and curve fitting, vol. 1 (2015)
9. Lu, S.: Accurate and efficient optic disk detection and segmentation by a circular transformation. IEEE Trans. Med. Imaging **30**(12), 2126–2133 (2011)

10. Airouche, M., Bentabet, L., Zelmat, M.: Image segmentation using active contour model and level set method applied to detect oil spills. In: Proceedings of the World Congress on Engineering, vol. 1, no. 1, pp. 1–3 (2009)
11. Sevastopolsky, A.: Optic disc and cup segmentation methods for glaucoma detection with modification of U-Net convolutional neural network. Pattern Recogn. Image Anal. **27**(3), 618–624 (2017)
12. Ronneberger, O., Fischer, P., Brox, T.: U-Net: convolutional networks for biomedical image segmentation. In: Navab, N., Hornegger, J., Wells, W.M., Frangi, A.F. (eds.) MICCAI 2015. LNCS, vol. 9351, pp. 234–241. Springer, Cham (2015). https://doi.org/10.1007/978-3-319-24574-4_28
13. Sivaswamy, J., Krishnadas, S., Joshi, G.D., Jain, M., Tabish, A.U.S.: Drishti-GS: retinal image dataset for optic nerve head (ONH) segmentation. In: 2014 IEEE 11th International Symposium on Biomedical Imaging (ISBI), pp. 53–56. IEEE (2014)
14. Vision 2020: The Right to Sight, IABP, Global-facts. http://www.iapb.org/vision-2020/what-is-avoidable-blindness/glaucoma
15. Lim, G., Cheng, Y., Hsu, W., Lee, M.L.: Integrated optic disc and cup segmentation with deep learning. In: 2015 IEEE 27th International Conference on Tools with Artificial Intelligence (ICTAI), pp. 162–169. IEEE (2015)
16. Sivaswamy, J., et al.: A comprehensive retinal image dataset for the assessment of glaucoma from the optic nerve head analysis. JSM Biomed. Imaging Data Pap. **2**(1), 1004 (2015)
17. Ronneberger, O., Fischer, P., Brox, T.: U-Net: convolutional networks for biomedical image segmentation. Computer Vision and Pattern Recognition (cs.CV), MICCAI (2015). arXiv:1505.04597 [cs.CV]
18. Abdel-Ghafar, R.A., Morris, T.: Progress towards automated detection and characterization of the optic disc in glaucoma and diabetic retinopathy. Med. Inform. Internet Med. **32**(1), 19–25 (2007). https://doi.org/10.1080/14639230601095865
19. Welfer, D., Scharcanski, J., Kitamura, C.M., Dal Pizzol, M.M., Marinho, D.R.: Segmentation of the optic disk in color eye fundus images using an adaptive morphological approach. Comput. Biol. Med. **40**(2), 124–137 (2010)
20. Maninis, K.-K., Pont-Tuset, J., Arbeláez, P., Van Gool, L.: Deep retinal image understanding. In: Ourselin, S., Joskowicz, L., Sabuncu, M.R., Unal, G., Wells, W. (eds.) MICCAI 2016. LNCS, vol. 9901, pp. 140–148. Springer, Cham (2016). https://doi.org/10.1007/978-3-319-46723-8_17
21. Zilly, J., Buhmann, J.M., Mahapatra, D.: Glaucoma detection using entropy sampling and ensemble learning for automatic optic cup and disc segmentation. Comput. Med. Imaging Graph. **55**, 28–41 (2017)
22. Walter, T., Klein, J.-C., Massin, P., Erginay, A.: A contribution of image processing to the diagnosis of diabetic retinopathy-detection of exudates in color fundus images of the human retina. IEEE Trans. Med. Imaging **21**, 1236–1243 (2002). https://doi.org/10.1109/TMI.2002.806290
23. Morales, S., Naranjo, V., Angulo, J., Alcañiz, M.: Automatic detection of optic disc based on PCA and mathematical morphology. IEEE Trans. Med. Imaging **32**, 786–796 (2013). https://doi.org/10.1109/TMI.2013.2238244
24. Abdullah, M., Fraz, M.M., Barman, S.A.: Localization and segmentation of optic disc in retinal images using circular hough transform and grow-cut algorithm. PeerJ **4**, e2003 (2006). Ed. Henkjan Huisman
25. Lowe, D.: Distinctive image features from scale-invariant keypoints. Int. J. Comput. Vis. **60**(2), 91–110 (2004)

26. Satpathy, A., Jiang, X., Eng, H.: Extended histogram of gradients feature for human detection. In: 2010 IEEE International Conference on Image Processing, Hong Kong, pp. 3473–3476 (2010). https://doi.org/10.1109/ICIP.2010.5650070
27. Liu, C., Yuen, J., Torralba, A.: SIFT Flow: dense correspondence across scenes and its applications. IEEE Trans. Pattern Anal. Mach. Intell. **33**(5), 978–994 (2011). https://doi.org/10.1109/TPAMI.2010.147
28. Chung, J.K., Kannappan, P.L., Ng, C.T., Sahoo, P.K.: Measures of distance between probability distributions. J. Math. Anal. Appl. **138**(1), 280–292 (1989)

Bird Species Classification Using Transfer Learning with Multistage Training

Akash Kumar[1]([✉]) and Sourya Dipta Das[2]([✉])

[1] Delhi Technological University, Delhi, India
akash_bt2k15@dtu.ac.in
[2] Jadavpur University, Kolkata, India
dipta.juetce@gmail.com

Abstract. Bird species classification has received more and more attention in the field of computer vision, for its promising applications in biology and environmental studies. Recognizing bird species are difficult due to the challenges of discriminative region localization and fine-grained feature learning. In this paper, we have introduced a Transfer learning based method with multistage training. We have used both Pre-Trained Mask-RCNN and a ensemble model consists of Inception Nets (InceptionV3 net & InceptionResnetV2) to get both the localization and species of the bird from the images. we have tested our model in an Indian bird dataset consist of variable size, high-resolution images are taken from camera in various environments (like day, noon, evening etc.) with different perspectives and occlusions. Our final model achieves an F1 score of 0.5567 or 55.67% on that dataset.

Code is available at: https://github.com/AKASH2907/bird-species -classification. Implemented in Keras [20].

Keywords: Bird species classification · Deep networks · Transfer learning · Multistage training · Object detection

1 Introduction

Bird species are recognized as useful biodiversity indicators. They are responsive to changes in sensitive ecosystems, whilst population-level changes in behavior are both visible and quantifiable. Suffered from great species variation, it is difficult for non-professionals to identify the sub-category of a bird only by its appearance. However, it is exhausting to annotate all the images by human beings with expert knowledge. Thus, an automatic classification system for bird species are needed, which will be a great convenience for many practical applications. For researchers working outdoors, shoot photos can be classified and analyzed immediately by the system, illustrated books are no more needed. For the public, the system could provide much fun when combined with culture information like poems and legends. It will arouse people's interest in birds and could benefit

---

A. Kumar and S. D. Das—Equal Contribution.

© Springer Nature Singapore Pte Ltd. 2019
C. Arora and K. Mitra (Eds.): WCVA 2018, CCIS 1019, pp. 28–38, 2019.
https://doi.org/10.1007/978-981-15-1387-9_3

the protection of birds. Apart from that, classifying bird species is an interesting problem for Fine-grained categorization, also known as subcategory recognition, which is a sub field in object recognition. In recent years, fine-grained classification stood out from basic-level classification, bringing promising applications and new challenges to computer vision society.

In this Bird Species Classification Challenge, our main focus was to classify birds from high-resolution photographs taken from a camera. In this task, to improve classification task, we have also provided localization of birds in the respective images with their class labels. The Main Challenges involving in this problem are given below.

1. Large Intensity variation in images as pictures are taken in different time of a day (like morning, noon, evening etc.)
2. Various poses of Bird (like flying, sitting with different orientation)
3. Bird localization in the image as there are some images in which there are more than one bird in that image
4. Large Variation in Background of the images
5. Various type of occlusions of birds in the images due to leaf or branches of the tree
6. Percentage of the object (bird) area in the image
7. Less number of sample images per class and also class imbalance.

We have proposed an end to end deep learning based approach using transfer learning to learn both micro and macro level features from Bird ROIs. We have used pre-trained Mask-RCNN to get the Bird ROIs from the images and used Multistage training method to remove class imbalance partially & boost up the accuracy of our model. Further information about the model is given in the respective section.

2 Related Work

In recent years, there is a number of existing works to automate the classification of the bird using audio data rather images. For this purpose, Feature Extraction from audio signals has some advantages like species have distinctive calls and no line of sight is needed for detection. However, there are some disadvantages also like an individual bird may emit no audio at all for an extended period of time and can't able to count the no of birds precisely. Due to this reasons, there are growing number of studies to use computer vision and image-based techniques for this problem [2,4,5]. Atanbori et al. [4] proposed a method to use motion features including curvature and wing beat frequency. Combined with Normal Bayes classifier and a Support Vector Machine classifier. Pang et al. [2] introduced discriminative features for bird species classification based on parts of birds and Marini et al. [5] proposed an approach to use a color segmentation to eliminate background elements and compute normalized color histograms to extract a feature vector for classification. Bird Species Classification with visual data is also important in the domain of fine-grained classification and there is a significant contribution to bird species classification problem respect to this domain [1,3,6,7]. There are also some few shots and deep learning based approaches [1,6,13] regarding this problem which has achieved a considered amount of accuracy in their respective datasets. These deep learning models failed miserably on our dataset. Therefore, we provide a robust solution to the area of classifying bird species in high-resolution images.

3   Dataset

In this paper, we have used the dataset from the CVIP 2018 Bird Species challenge. Training dataset consists of 150 images with 16 species of birds and testing dataset contains 158 images. The dataset contains high resolution images ranging from 800 × 600 to 4000 × 6000. The dataset is class imbalanced with 5 images in one species to 20 images in other species. Those bird images were taken in different environments, occlusions, perspectives. Due to those situations, it becomes really hard to recognize bird from the images without their localization. Overview of the dataset is shown in Fig. 1 below.

Fig. 1. Dataset overview

4   Proposed Approach

4.1   Data Augmentation

To increase the number of training samples per class and reduce the effect of class imbalance, data augmentation is used. Relevant image augmentation techniques are chosen according to the bird type of each class. Those techniques are Gaussian Noise, Gaussian Blur, Flip, Contrast, Hue, Add (add some values to each channel of the pixel), Multiply (multiply some values to each channel of the pixel), Sharp, Affine transform. As the dataset was quite small, the networks trained on the dataset, overfitted the dataset and does not generalize well on 150 images. After data augmentation, training dataset increased from 150 images to 1330 images. Some of the data augmentations were not used for some classes to reduce the effects of class imbalance. Augmentation techniques per class were chosen according to the context and practical viability of the birds in that environment. More details on those augmentation techniques per class were mentioned in the table below (Table 1).

Table 1. Table for data augmentation techniques for each of bird species

Species	Gaussian Noise	Gaussian Blur	Flip	Contrast	Hue	Add	Multiply	Sharp	Affine	Total
blasti	Yes	Yes	Yes	Yes	Yes	No	No	No	No	90
bonegl	Yes	Yes	Yes	Yes	Yes	Yes	Yes	Yes	Yes	78
brhkyt	Yes	Yes	Yes	Yes	Yes	Yes	Yes	Yes	Yes	65
cbrtsh	Yes	Yes	Yes	Yes	Yes	Yes	Yes	Yes	Yes	91
cmnmyn	Yes	Yes	Yes	Yes	Yes	Yes	Yes	Yes	Yes	91
gretit	Yes	Yes	Yes	Yes	Yes	Yes	Yes	Yes	Yes	78
hilpig	Yes	Yes	Yes	Yes	Yes	Yes	Yes	No	No	80
himbul	Yes	Yes	Yes	Yes	Yes	No	No	No	No	99
himgri	Yes	Yes	Yes	Yes	Yes	No	No	No	No	100
hsparo	Yes	Yes	Yes	Yes	Yes	No	No	No	No	81
indvul	Yes	Yes	Yes	Yes	Yes	No	No	No	No	81
jglowl	Yes	Yes	Yes	Yes	Yes	Yes	Yes	Yes	Yes	78
lbicrw	Yes	Yes	Yes	Yes	Yes	Yes	Yes	Yes	Yes	78
mgprob	Yes	Yes	Yes	Yes	Yes	Yes	Yes	Yes	Yes	78
rebimg	Yes	Yes	Yes	Yes	Yes	Yes	Yes	No	No	80
wcrsrt	Yes	Yes	Yes	Yes	Yes	Yes	Yes	No	No	80

4.2   Bird ROI (Region of Interest) Detection

To eliminate background elements or regions and also extract features from specific bird class, pretrained Object Detection deep nets are used. In this Model, we have used Mask R-CNN [8] to localize birds in each image from both test & training dataset. We have used the pretrained weights of Mask R-CNN, trained on the COCO dataset [14] which contains 1.5 million object instances with 80 object categories (including birds) (Fig. 2).

Mask R-CNN [8] comprises two architectures: Faster R-CNN [17] and Fully Connected Network [18]. The details of the two architectures are summarized as below:

(1) Faster R-CNN [17] runs a parallel network with two heads: Classification head and Bounding Box Regression head. Faster R-CNN also uses a small convolutional network as Region Proposal Network (RPN) to refine the region of interest.

(2) Fully Connected Networks (FCN) [18] is a meta-algorithm used for semantic segmentation process. It is built from convolutional, pooling, downsampling and upsampling layers. Due to absence of dense layers in FCN, the number of parameters are less and hence computation time reduces drastically. Due to the inclusion of pixel level info in ROI alignment, it works better in comparison to ROI pooling that is used in Faster R-CNN.

4.3 Transfer Learning

In our case, transfer learning learns both micro and macro level feature extracted from bird images for classification. We have used ImageNet [9] pretrained weights to initialize our Deepnet model for training. ImageNet contains 1.2 million images belonging to 1000 classes. Training using pretrained ImageNet weights help us to learn fine-grained as well as global level features beforehand and learn the deepnet more specific & discriminative features for each bird species which leads to increase the accuracy of our model.

In our task, we used Inception V3 [11] and Inception ResNet V2 [10] for finetuning. We used VGG-16/19 [16] initially, but the classification accuracy was very low. Then, we moved on to ResNet [15] and Inception V3 [11] that have almost similar accuracy on ImageNet dataset, in our case Inception V3 performed better. Inception ResNet V2 is derived from the base architectures of Inception and ResNet. It has good accuracy as well as less computation time.

Inception V3 have features like Batch Normalization, Factorization and varied size kernels that helps to learn the global and local features. Also, the computation time reduces well due to the factorization method. Inception ResNet V2 incorporates the backbone from Inception and ResNet modules. From ResNet, skip connections were introduced that improved the accuracy by retaining features deep into the layers. In Inception modules, skip connections were introduced alongwith Factorization to design Inception ResNet V2 architecture. It reduced the computation time as well as helped in the propagation of all the features deeper into the layers.

Fig. 2. Mask R-CNN cropped birds images

4.4   Ensemble Model Architecture

We have used InceptionResNetV2 [10] & InceptionV3 [11] deepnet architectures to create a ensemble model as our classification model. The prediction vector from Inception V3 [11] and Inception ResNet V2 [10] weights are generated for each image at the time of testing. There are two cases with Mask R-CNN:

(1) Birds Detection: If the Mask R-CNN detect birds in the image, then a batch of cropped bird images are created. The whole batch for that particular image is evaluated using both network weights. Both the prediction vectors are compared and then the species is assigned based on the prediction value with the highest weight or prediction confidence value of the species of bird is finally predicted.
(2) No Bird Detection: The original image is predicted for the species of the bird using both the architecture weights. Though, the number of such cases is very less. The species with the highest predicted value is added to the final prediction vector.

Figure 3 illustrates the overall process of bird detection and species classification using Mask R-CNN and ImageNet models respectively.

5   Experiments

5.1   Multi-stage Training

We used multi-stage training to improve the accuracy of the model. Firstly, we trained the Inception V3 architecture and then Inception ResNet V2 architecture on data augmented original images. In the second stage, we used the pretrained weight on original images to train on cropped images generated from Mask R-CNN. All the images are resized to 416 × 416. The accuracy of the model increased by 2–3 % after training on the cropped images. The multi-stage training helps to learn fine-grained features using cropped images of birds and the original images are used to learn the global spatial features present in the image.

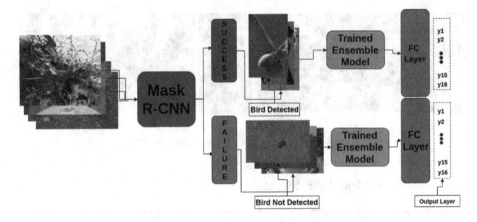

Fig. 3. Overall architecture of our bird species classification system

5.2   Testing

At the time of testing, the images are passed from Mask R-CNN pre-loaded with COCO weights. The cases of Mask R-CNN is discussed in the Ensemble Model Architecture. Here, we have used 'categorical cross-entropy' as loss function & Adam as an Optimizer to train both deep networks. For fine-tuning of the deep-net models, we have tested on various types of activation functions and Swish activation function [12] performed best among all of them.

6   Evaluation

Evaluation Metric. For the challenge, we used three evaluation metrics -

1. Precision: It is the ratio of correctly predicted positive observations to the total observations. It is defined as:

$$\frac{TruePositives}{TruePositives + FalsePositives} \tag{1}$$

2. Recall: It is the ratio of correctly predicted positive observations to all the observations in the relevant class.

$$\frac{TruePositives}{TruePositives + FalseNegatives} \tag{2}$$

3. F1-score: It is the harmonic mean of precision and recall.

$$2 * \frac{precision * recall}{precision + recall} \tag{3}$$

The evaluation metrics are calculated from the Confusion matrix. In confusion matrix, True positives is equal to sum of diagonal elements. False positives is equal to the sum of each column excluding diagonal elements. False negatives is equal to the sum of each row elements excluding diagonal elements.

7   Results

Table 2 contains the F1-scores obtained from different architectures trained on original images and Mask R-CNN crops as discussed in Multi-Stage Training. The Inception ResNet V2 model trained firstly on resized original images and then on Mask R-CNN cropped images gives the best results at the time of training. The final trained weights of this model was used to predict the specie of bird.

Table 2. Accuracy during Multi-Stage Training on InceptionResnetV2 & InceptionV3 models.

Model architecture	Data subset	Train	Validation	Test
Inception V3	Images	91.26	12.76	30.95
	Images + Crops	93.97	15.50	41.66
Inception Resnet V2	Images	**97.29**	29.17	47.96
	Images + Crops	92.29	**33.69**	**49.09**

Table 3 summarizes the class averaged precision, recall, and F1 scores. In the ensemble method, the prediction vector for both Inception V3 and Inception Resnet V2 is compared for each class predicted. The network with a higher probability of a particular species was appended to the final prediction file.

Table 3. Evaluation metrics (in %) on test dataset

Model architecture	Precision	Recall	F1
Mask R-CNN + InceptionV3	48.61	45.65	47.09
Mask R-CNN + InceptionResnetV2	53.62	48.72	51.05
Mask R-CNN + Ensemble Model	**56.58**	**54.8**	**55.67**

We also used the confusion matrix to get more intuition about the performance of different ImageNet architectures. The confusion matrix for final architecture is shown in Fig. 4. From the confusion matrix, we can see that our approach failed miserably in four classes particularly.

In order to study the limitations of our architecture, we also analyzed our approach on Caltech-UCSD Birds-200-2011 [19] dataset. The main problem we faced that the images in this dataset are very different from our dataset. The images are well-uniform and in every image, birds at least cover 90% of the image. The images in this dataset contain only one instance of bird species per image. Hence, the addition of the Mask R-CNN to the ImageNet pipeline does not provide any substantial additional gains.

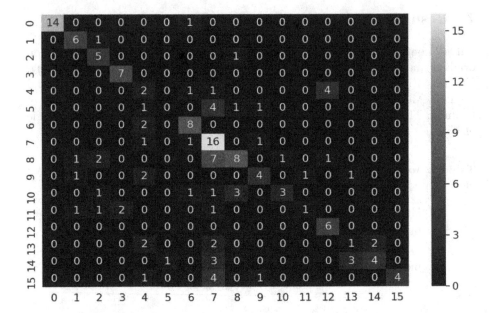

	0	1	2	3	4	5	6	7	8	9	10	11	12	13	14	15
0	14	0	0	0	0	0	1	0	0	0	0	0	0	0	0	0
1	0	6	1	0	0	0	0	0	0	0	0	0	0	0	0	0
2	0	0	5	0	0	0	0	0	1	0	0	0	0	0	0	0
3	0	0	0	7	0	0	0	0	0	0	0	0	0	0	0	0
4	0	0	0	0	2	0	1	1	0	0	0	0	4	0	0	0
5	0	0	0	0	1	0	0	4	1	1	0	0	0	0	0	0
6	0	0	0	0	2	0	8	0	0	0	0	0	0	0	0	0
7	0	0	0	0	1	0	1	16	0	1	0	0	0	0	0	0
8	0	1	2	0	0	0	0	7	8	0	1	0	1	0	0	0
9	0	1	0	0	2	0	0	0	0	4	0	1	0	1	0	0
10	0	0	1	0	0	0	1	1	3	0	3	0	0	0	0	0
11	0	1	1	2	0	0	0	1	0	0	0	1	0	0	0	0
12	0	0	0	0	0	0	0	0	0	0	0	0	6	0	0	0
13	0	0	0	0	2	0	0	2	0	0	0	0	0	1	2	0
14	0	0	0	0	0	1	0	3	0	0	0	0	0	3	4	0
15	0	0	0	0	1	0	0	4	0	1	0	0	0	0	0	4

Fig. 4. Confusion matrix of Mask R-CNN + Ensemble model

8 Error Analysis

From the confusion matrix, cmnmyn, gretit, hsparo, indvul, mgprob, and rebimg classes have poor accuracy. Few training samples per class with large variance in the object (bird) size in the image and class imbalance are the main reasons for those misclassifications. Though there are some other reasons like small bird ROIs, the similarity in bird body part's color & background color, occlusion, and variation in lighting condition in training and testing dataset. For those conditions, deep nets can't be able to learn those discriminative features (both micro features like color, gradients, textures etc. and macro features like shape, color patch etc.). Various lighting condition (like a picture taken during daylight, dawn, dusk, evening etc.) affected our model most as due to low background light, many micro-features of the bird like color, texture, gradients etc. are lost. Different poses of bird also reduced our model's accuracy though it is compensated with micro-level features of the bird. A possible solution for those problems is discussed further in the later section (Fig. 5).

Fig. 5. Few examples of images where Mask R-CNN fails to detect birds

9   Conclusion and Future Work

In this paper, we have proposed a method to both localize and classify the species of the bird from high-definition photographs taken from a camera by using an end-to-end approach with Mask R-CNN, transfer learning, and multistage training. We have considered this challenge and as a few shot classification problem and proposed to use bird localization in images to boost up the accuracy of our model. Transfer learning helped our model to learn more specific and converge loss function more quickly with good accuracy on the test dataset. In future, we are planning to extend this work using a Part Model-based approach with NTM (Neural Turing Machine) and Visual Attention Network.

References

1. Fu, J., Zheng, H., Mei, T.: Look closer to see better: recurrent attention convolutional neural network for fine-grained image recognition. In: 2017 IEEE Conference on Computer Vision and Pattern Recognition (CVPR), Honolulu, HI, pp. 4476–4484 (2017). https://doi.org/10.1109/CVPR.2017.476
2. Pang, C., Yao, H., Sun, X.: Discriminative features for bird species classification. In: Proceedings of International Conference on Internet Multimedia Computing and Service, ICIMCS 2014, p. 256, 5 p. ACM, New York (2014). https://doi.org/10.1145/2632856.2632917
3. Ge, Z., McCool, C., Sanderson, C., Bewley, A., Chen, Z., Corke, P.: Fine-grained bird species recognition via hierarchical subset learning. In: 2015 IEEE International Conference on Image Processing (ICIP), Quebec City, QC, pp. 561–565 (2015). https://doi.org/10.1109/ICIP.2015.7350861
4. Atanbori, J., Duan, W., Murray, J., Appiah, K., Dickinson, P.: Automatic classification of flying bird species using computer vision techniques. Pattern Recogn. Lett. **81**, 53–62 (2016). https://doi.org/10.1016/j.patrec.2015.08.015
5. Marini, A., Facon, J., Koerich, A.: Bird species classification based on color features. In: Proceedings - 2013 IEEE International Conference on Systems, Man, and Cybernetics, SMC 2013, pp. 4336–4341 (2013). https://doi.org/10.1109/SMC.2013.740
6. Branson, S., et al.: Bird species categorization using pose normalized deep convolutional Nets. CoRR abs/1406.2952 (2014)
7. Yang, S., et al.: Unsupervised template learning for fine-grained object recognition. In: NIPS (2012)
8. He, K., et al.: Mask R-CNN. In: 2017 IEEE International Conference on Computer Vision (ICCV), pp. 2980–2988 (2017)
9. Krizhevsky, A., Sutskever, I., Hinton, G.E.: ImageNet classification with deep convolutional neural networks. In: NIPS (2012)
10. Szegedy, C., et al.: Inception-v4, Inception-ResNet and the impact of residual connections on learning. In: AAAI (2017)
11. Szegedy, C., Vanhoucke, V., Ioffe, S., Shlens, J., Wojna, Z: Rethinking the inception architecture for computer vision. In: 2016 IEEE Conference on Computer Vision and Pattern Recognition (CVPR), Las Vegas, NV, United States, pp. 2818–2826 (2016). https://doi.org/10.1109/CVPR.2016.308
12. Ramachandran, P., Zoph, B., Le, Q.V.: Swish: a self-gated activation function (2017)

13. Vinyals, O., Blundell, C., Lillicrap, T.P., Kavukcuoglu, K., Wierstra, D.: Matching networks for one shot learning. In: NIPS (2016)
14. Lin, T.-Y., et al.: Microsoft COCO: common objects in context. In: Fleet, D., Pajdla, T., Schiele, B., Tuytelaars, T. (eds.) ECCV 2014. LNCS, vol. 8693, pp. 740–755. Springer, Cham (2014). https://doi.org/10.1007/978-3-319-10602-1_48
15. He, K., Zhang, X., Ren, S., Sun, J.: Deep residual learning for image recognition. CoRR, abs/1512.03385 (2015)
16. Simonyan, K., Zisserman, A.: Very deep convolutional networks for large-scale image recognition. CoRR, abs/1409.1556 (2014)
17. Ren, S., He, K., Girshick, R., Sun, J.: Faster R-CNN: towards real-time object detection with region proposal networks. In: Proceedings of the 28th International Conference on Neural Information Processing Systems - Volume 1, NIPS 2015, Montreal, Canada (2015)
18. Shelhamer, E., Long, J., Darrell, T.: Fully convolutional networks for semantic segmentation. IEEE Trans. Pattern Anal. Mach. Intell. **39**, 640–651 (2017)
19. Wah, C., Branson, S., Welinder, P., Perona, P., Belongie, S.: The Caltech-UCSD Birds-200-2011 dataset. Computation & Neural Systems. Technical Report, CNS-TR-2011-001 (2011)
20. Chollet, F., et al.: Keras (2015). https://keras.io

A Deep Learning Paradigm
for Automated Face Attendance

Rahul Kumar Gupta, Shreeja Lakhlani, Zahabiya Khedawala,
Vishal Chudasama, and Kishor P. Upla[✉]

Sardar Vallabhbhai National Institute of Technology, Surat, Gujarat, India
singhalrahul222@gmail.com, shreejald@gmail.com,
zahabiyakhedawala@gmail.com, vishalchudasama2188@gmail.com,
kishorupla@gmail.com

Abstract. In this paper, we propose an end-to-end automatic face attendance system using Convolutional Neural Networks (CNNs). Attendance of a student plays an important role in any academic organization. Manual attendance system is very time consuming and tedious. On the other hand, automatic attendance system through face recognition using CCTV camera can be fast and can reduce the man-power involved in that process. Here, we have pipelined one of the best existing architectures such as: (i) Single Image Super-Resolution Network (SRNet) for image super-resolution, (ii) MTCNN for face detection and (iii) FaceNet for face recognition in order to come up with a novel idea of marking attendance. Due to poor video quality of CCTV camera, it becomes difficult to detect and recognize faces accurately and this may reduce the attendance accuracy. To overcome this limitation, we propose a CNN framework called SRNet which super-resolves a given low resolution (LR) image and also increases the face recognition accuracy. We make use of five different datasets i.e. RAISE and DIV2K for SRNet, VGGface2 for FaceNet, LFW and our own dataset for testing and validation purpose. The proposed face attendance system displays a sheet which consists of a list of absent and present persons and the overall attendance record. Our experimental results show that the proposed approach outperforms other existing face attendance approaches.

Keywords: Deep learning · Convolutional Neural Networks · SRNet ·
MTCNN · FaceNet · Face attendance

1 Introduction

Advances in the field of face detection and recognition using CNN can be exploited to eliminate human intervention in marking attendance. Face recognition of images captured by a CCTV camera eliminates manual efforts, saves time and is a cost effective solution. But many a times due to low spatial resolution of CCTV video footage, some students are not detected and recognized which results in absentees in the list. Making a robust automatic face attendance system for LR image from a CCTV camera could be an active area of research.

© Springer Nature Singapore Pte Ltd. 2019
C. Arora and K. Mitra (Eds.): WCVA 2018, CCIS 1019, pp. 39–50, 2019.
https://doi.org/10.1007/978-981-15-1387-9_4

In the proposed approach, we make use of a pipe-lined architecture based on CNN to build an end-to-end face attendance system. We capture a video frame from a CCTV camera. Now, if its spatial resolution is poor then it is passed through the proposed SRNet module. SRNet is trained using MSE based loss function which results in SR images with better Peak Signal-to-Noise Ratio (PSNR) and Structural Similarity (SSIM) measures in addition to preservation of high frequency details in the image. The SR image is then fed to a Multi-Mask Cascaded Neural Network (MTCNN) [30] that detects faces in the image in a coarse-to-fine manner. It also results in five essential facial landmark positions. The detected faces are cropped around the bounding boxes, aligned and then given to FaceNet [21], a face recognition module which is used to identify those faces. Finally, recognized faces are marked in attendance sheet. Following are the main contributions of the proposed face attendance system:

- The proposed automatic face attendance system using CNNs has a multi-task and a multi-frame architecture which gives better accuracy as compared to other existing face attendance methods.
- Performance of the proposed SRNet shows significant improvement as compared to other state-of-the-art methods in terms of PSNR and SSIM.

2   Related Work

CNN Based Single Image Super-Resolution (SISR) Methods: Dong et al. [5] propose three convolutional layered end-to-end model called SRCNN in which a bicubic interpolation is used at first to upsample the given LR image. Thereafter, the very deep convolutional network (i.e., VDSR) [9] is proposed by increasing the depth of network as 20 convolutional layers. In VDSR, the global residual learning paradigm is adopted to generate the final residual image. Deep recursive convolutional network (i.e., DRCN) is proposed by Kim et al. [10] where 16 recursive layers are used to keep a small number of model parameters. DRCN obtains better PSNR value than that of obtained using SRCNN and VDSR methods. Tai et al. [25] propose deep recursive residual network (i.e., DRRN) and extend the local residual learning approach [7] with 52-layered deep network. Recently, Lai et al. [12] introduce progressive reconstruction approach for SISR and propose a model called LapSRN in which the SR images are progressively reconstructed at multiple pyramid levels. In LapSRN, the more robust Charbonnier loss function [2] is used to train their network and obtain better SR images. Recently, Ledig et al. [13] propose a SR method called super-resolution using residual network (i.e., SRResNet) which sets a new state-of-the-art performance in SISR.

CNN Based Face Detection and Recognition: Pioneer work in face detection is carried out by Viola and Jones [27]. They propose a face cascade detector which learns a rich set of image features using Haar-classifier and yields an extremely efficient classifier based on the AdaBoost algorithm. This detector works well in real-time but its performance degrades with large appearance variance. Yang et al. [29] train the CNN to discover facial parts responses from arbitrary uncropped face images. Li et al. [14] introduce a CNN-based face bounding box calibration step in the cascade form to obtain a high quality localization. Multi-Task Cascaded Convolution Network (MTCNN) [30] use a deep cascaded multi-task framework to exploit the inherent correlation between detection and alignment to boost up their performance.

Taigman et al. [26] propose a facial alignment system based on explicit 3D modelling of faces. Sun et al. [23] perform face recognition using two supervisory signals simultaneously i.e. face identification and verification. Unlike 3D modelling, it is based on simple 2D affine transformation. To overcome the challenges of pose and illumination, FaceNet [21] presents a unified system for face verification, recognition and clustering.

Attendance System: Rathod et al. [19] propose a face attendance system based on Viola Jones [27] face detection by using HoG based features extraction and SVM classifier for recognition. Chintalapati et al. [3] develop an automated face attendance management system which use Viola Jones for face detection followed by histogram equalization for feature extraction using PCA and finally SVM for multi-class classification. These methods are based on conventional machine learning algorithms which have limited learning ability. University classroom attendance proposed by Fu et al. [6] integrate two deep learning algorithms - MTCNN for face detection and Center-Face [28] for recognition. Though their approach obtains high accuracy but it is only well suited for images from a high resolution camera.

3   Methodology

Figure 1 displays the block schematic representation of the proposed end-to-end automatic face attendance system which consists of the following five modules:

- Super-resolution module
- Face detection module
- Intermediate processing
- Face recognition module
- Attendance module

Super-Resolution Module: This module is used to reconstruct the HR image from the given LR observation captured using the CCTV camera. The architecture of the proposed SR method i.e. SRNet is displayed in Fig. 2 for $\times 4$ upscaling factor. 9×9 filters in the beginning of SRNet increase the size of the receptive field and help in extracting prominent features. Inspired by VGG architecture [22], remaining filters are 3×3 to create a deeper model with less number of the parameters to be trained. $M = 16$ number of residual blocks are used. In order to up-sample the feature map in SRNet by $\times 4$, we use two resize-conv's where each of them up-samples by a factor of 2. We use the residual network suggested

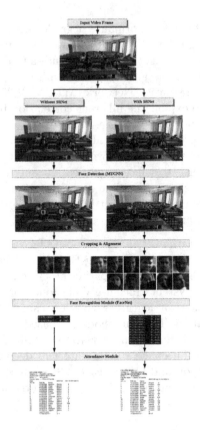

Fig. 1. Block schematic representation of the proposed end-to-end automatic face attendance system.

by Lim et al. [15] in which batch normalization (BN) layers are removed from original residual network proposed in [7]. This helps to reduce the GPU memory [15]. As shown in Fig. 2, the local residual learning (LRL) is used in the residual network as shown in Fig. 2. This LRL helps to solve the problem of exploding or vanishing gradient since the higher layer gradients are directly passed to the lower layer in residual block. We also adopt global residual learning (GRL) as suggested by Kim et al. [9] in which the model's output is added with the bicubic interpolation of the input image to generate the residual image. Such GRL helps the network to learn the identity function for LR test image and stabilze the training process and also reduces the color shifts in output image. However, we modify the GRL network by passing the bicubically interpolated image through two convolution layers which helps us to extract more useful features of LR image further (see Fig. 2). Instead of using transpose layer [16] to upsample the feature maps inside the network, we use resize convolution (i.e., nearest-neighbor interpolation followed by a single convolution layer) as suggested by Odena et al. [17]. The use of such upsampling layer reduces the checkerboard artifacts in

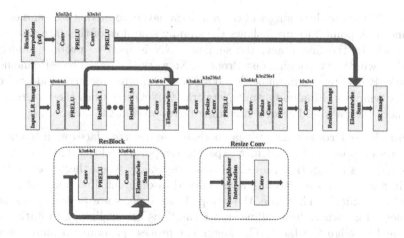

Fig. 2. Block diagram of proposed super-resolution network: SRNet. Here, k, s and n indicate the size of filters, stride value and number of filters in feature map respectively and $M = 16$ in the network.

Fig. 3. Cascaded framework for face detection using MTCNN [30].

the SR image. However, this approach still creates some checkerboard artifacts for some specific loss functions [20]. Hence, we add two convolution layers after resize convolution layer in the proposed method which can work as an additional regularization term to reduce the checkerboard artifacts further.

The most commonly used pixel-wise MSE loss function is used to train the proposed SRNet model. The MSE loss function between HR image, I^{HR} and super-resolved image I^{SR} is defined as,

$$l_{MSE} = \frac{1}{r^2wh} \sum_{x=1}^{rw} \sum_{y=1}^{rh} (I_{x,y}^{HR} - I_{x,y}^{SR})^2. \tag{1}$$

Here, r is the downsampling factor while h and w is denoted as the height and width of the LR observation, respectively. By minimizing the MSE loss function, we can obtain high PSNR value. Hence, this MSE loss function is used in most of the state-of-the-art SISR methods to obtain better PSNR values [4,5,9,10].

Face Detection Module: After the reconstruction of SR image using SRnet, the next step is to detect the present faces in that image. Hence, we use state-of-the-art face detection module called Multi-Task Cascaded Neural Network (MTCNN) [30]. This framework uses a cascaded architecture for multi-task learning. There are three stages of convolutional layers to predict face and essential landmark points. Figure 3 shows the step-by-step flow of this face detection framework. In the first stage, the shallow CNN is used to produce multiple candidate windows through a fast Proposal Network (P-Net). Then, Refinement Network (R-net) results in refined windows by discriminating face windows from non-face windows. Finally, Output Network (O-Net) identifies the face regions with more accuracy and creates bounding boxes around the faces.

Intermediate Processing: In this module, the detected faces with bounding boxes are cropped and saved in a separate temporary folder. This step helps in making face recognition procedure easier especially when there are multiple faces in a single image. Faces that are turned in different directions make face recognition difficult. Therefore, the cropped face images are passed to the face alignment stage where basic affine transformations like scaling and rotation are performed to align the faces. The alignment process generates a more tightly cropped image and removes the unwanted background.

Face Recognition Module: The aligned face images are applied to face recognition module. Here, we employ FaceNet [21], a face recognition framework, which is based on the Inception-Resnet v1 model [24]. This model generates a 512-dimensional Euclidean embeddings for each face during training process. To train the deep neural network, FaceNet employs Triplet loss function. The loss function ensures that an image x_i^a (anchor) of a specific person is closer to all other images x_i^p (positive) of the same person than it is to any image x_i^n (negative) of any other person. Hence the loss function is given by [21],

$$L = \sum_{i}^{N}[||f(x_i^a) - f(x_i^p)||_2^2 - ||f(x_i^a) - f(x_i^n)||_2^2 + \alpha], \tag{2}$$

where α is the margin between positive and negative pairs and f(x) is the generated embedding. The face embeddings are used as a measurement of face similarity. Pairs of faces of the same person with different pose and illumination have less output distance between them whereas faces of distinct people have large distances.

Face Attendance Module: In this module, face attendance sheet is created based on the recognized faces. Before the attendance is marked, the faculty is prompted to enter his name, the name of the subject and the date for which he/she wants to mark the attendance. These details get updated in the attendance sheet. Our database consists of the three folders based on the courses offered by the institute: undergraduate (UG), post graduate (PG) and PhD. The database contains a list of students enrolled in a particular course. Month-wise and subject-wise attendance is marked in the excel sheet according to the roll numbers present during each lecture. The identified faces are recognized using their roll numbers. The roll numbers follow a naming convention (for example P17EC003). In the roll number, the first letter gives the information of course name, the next two numbers give details of admission year, the third and fourth letters show the name of department and the last three numbers signify the serial number of particular student. We have a four tier directory structure as,

```
Course Name
 └─ Admission Year
     └─ Department
         └─ Faculty Name
```

From the roll number of the recognized student, we extract the path to the attendance sheet and the attendance of the student is marked for the current date and subject.

4 Experimental Results

We have tested and validated the performance of the proposed end-to-end face attendance system on LFW dataset. However, the proposed method SRNet has been tested on Set5, Set14 and BSD100 benchmark datasets. Results of SRNet are compared with the online supplementary materials of other existing state-of-the-art methods such as SRCNN[1] [5], LapSRN[2] [12] and EnhanceNet[3] [20]. For quantitative comparison, the common evaluation metrics i.e., PSNR and SSIM are used and the same are calculated after converting SR images into YCbCr color space and removing the four boundary pixels of Y-channel images as suggested in [15, 20].

Training Details and Hyper-parameter Settings: All the modules in our pipelined architecture have been chosen with due care considering their pros and cons. These modules have their own advantages when used as individual blocks, however, in our proposed work we have pipelined them and we leverage their individual performance capabilities to build an end-to-end application based on face recognition. These modules are trained individually to optimize the performance of our end-to-end pipelined network.

To train the proposed method SRNet, we use two datasets: RAISE and DIV2K. Before the training process, images are augmented with flipping, random rotation (upto $270°$) and downscaling (by a factor of 0.5 to 0.7 randomly) operations. The HR images are bicubically downsampled with a factor $r = 4$ to produce the corresponding LR images. Inspired from Kim et al. [9], we adopt a two-stage training strategy to train proposed SRNet model which helps to avoid undesired local minima. At first stage, the proposed SRNet is trained on RAISE dataset upto 8×10^5 number of iterations with a learning rate of 10^{-4} using MSE loss function. After this, the same model is further trained upto 4×10^5 number of additional iterations with same learning rate and loss function on DIV2K dataset. The Adam optimizer [11] with $\beta_1 = 0.9$ is used to optimize the SRNet model.

---

[1] https://github.com/jbhuang0604/SelfExSR.
[2] http://vllab.ucmerced.edu/wlai24/LapSRN/.
[3] http://webdav.tuebingen.mpg.de/pixel/enhancenet/.

For face detection task, we choose a pre-trained model of MTCNN[4] [30] framework which was trained on WIDER FACE dataset. This dataset consists 393,703 labelled face image with bounding boxes in 32,203 images. The MTCNN framework jointly performs the task of face detection and alignment and outperforms other state-of-the-art methods [29] [14] in terms of speed and validation accuracy. We use the face recognition module FaceNet[5] [21] which is trained on VGGFace2 dataset which consists of 3.31 million images of 9131 subjects. FaceNet trains its deep convolutional network based on an online Triplet based loss function. Here, we employ transfer learning approach [18] to further train FaceNet on our own dataset of students. In the proposed work, we have acquired 50 photos of every student and then same are used with image augmentation to transform each photo randomly into 10 different photos, so in total it consists 500 photos for each person. Since we have considered 16 students for this work, therefore our training dataset has 8000 images.

Table 1. Quantitative comparison in terms of PSNR and SSIM of the proposed SRNet with other existing CNN based SR methods. Here, highest measures are indicated in bold.

Method	PSNR			SSIM		
	Set5	Set14	BSD100	Set5	Set14	BSD100
Bicubic	28.4302	26.0913	25.9619	0.8109	0.7043	0.6675
SRCNN [5]	30.0843	27.2765	26.7046	0.8527	0.7425	0.7016
SelfExSR [8]	30.3485	27.5539	26.8531	0.8631	0.7551	0.112
VDSR [9]	31.3537	28.1101	27.2876	0.8839	0.7691	0.725
DRCN [10]	31.5405	28.1219	27.238	0.8855	0.7686	0.7232
LapSRN [12]	31.5417	28.1852	27.3175	0.8863	0.7706	0.7259
MS-LapSRN [12]	31.7368	28.3559	27.4241	0.8899	0.7749	0.73
EnhanceNet-E [20]	31.7568	28.4297	27.5112	0.8886	0.7771	0.7319
SRNet	**32.0916**	**28.5397**	**27.5212**	**0.8933**	**0.7795**	**0.7343**

Testing and Evaluation: Performance of the proposed SRNet model is quantitatively compared in terms of PSNR and SSIM measures in Table 1. In this table, we observe that SRNet has the highest PSNR and SSIM values as compared to other existing CNN based SR methods.

In order to see the importance of super-resolution task, the original image is downsampled by different factors like 2, 3 and 4 to reduce its spatial resolution. These images are then given to the face detection module with or without passing through SRNet. Figure 4a shows the results of detected faces for these two cases. By looking at the figure, one can notice that the face detection module with

---

[4] https://github.com/kpzhang93/MTCNN_face_detection_alignment.
[5] https://github.com/davidsandberg/facenet.

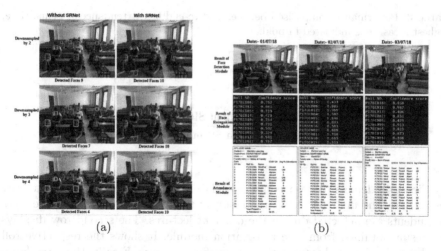

Fig. 4. (a) Face detection ability of SRNet with different downsampling factors. (b) Procedure of updating attendance sheet on daily basis.(Reader can zoom-in for better view.)

Table 2. Quantitative measurement of face detection module with and without SRNet on the images of WIDER FACE dataset.

	Original LR image W/o/with SRNet (%)	Downsampled by 2 W/o/with SRNet (%)	Downsampled by 3 W/o/with SRNet (%)
Precision	99.3/92.45	98.68/93.53	98.98/94.18
Recall	30.12/79.65	10.48/45.01	3.42/24.32
F1 Score	46.25/85.56	18.07/60.77	6.61/38.65
Accuracy	29.96/74.46	10.47/43.66	3.42/23.96

SRNet can detect 10 students out of 11 in each of the downsampled images while the face detection module without SRNet can detect only a few students when the image resolution becomes very poor, as seen in the last row of Fig. 4a. Thus, SRNet helps to detect faces even in very low spatial resolution images and poorly lit conditions.

We evaluate face detection module on WIDER FACE dataset. Total 292 group images consisting of 2861 faces from this dataset are first downsampled by a factor of 4 which form the original LR images. These are further downsampled by factor of 2 and 3 in order to evaluate our results on very low resolution images. These downsampled images are applied to the proposed SR module which results in SR images. The LR and the corresponding SR images are passed though face detection module and quantitative measurements such as precision, recall, F1 score and accuracy are calculated. Table 2 summarizes these quantitative measurements calculated from the results of face detection task on WIDER FACE dataset.

Table 3. Performance comparison between end-to-end face attendance systems. Here, highest measure is indicated in bold.

Model	Accuracy
Chintalapati et al. [3]	0.95
FaceTime [1]	0.9502
Proposed framework (without SR module)	0.9665
Proposed framework (with SR module)	**0.9680**

Figure 4b depicts the procedure of updating the attendance sheet on a daily basis. Here, the first row shows the result of face detection module where faces of students are detected during the different lectures. The second row displays the result obtained from face recognition module. It shows the respective roll numbers of recognized faces with their confidence score. Finally, the attendance sheet is prepared based on the recognized roll numbers (see third row in Fig. 4b).

We compare the performance of our proposed framework with the other existing end-to-end face attendance systems. Table 3 lists the accuracy obtained using the different existing end-to-end face attendance systems. We evaluated the performance of our proposed system on 6000 pairs of LFW faces and it could achieve an accuracy of 96.80%. Here, one can notice that the proposed framework obtains better accuracy value when compared to other existing frameworks. We also validated and tested our system on low quality images without super-resolution and it could obtain an accuracy of 96.65%.

5   Conclusion

In this paper, we propose a novel pipe-lined architecture for an end-to-end real time face attendance system based on super resolution and CNN based face detection and recognition. The proposed SRNet is a promising module in this work and is the main highlight that sets our work apart from other existing face attendance management systems [1,6,19,19]. Experimental results demonstrate that the proposed system works well even with low resolution images captured in real time by a CCTV camera and obtains an accuracy of 96.80% on the LFW benchmark dataset for evaluation. The marked improvement in the end-to-end system opens the door for implementing this system as a convenient biometric approach in any organization. Future work will focus on building a GUI application that can run on smartphones and can be accessed handily from anywhere. Hazy images due to motion blur can be enhanced by employing image deblurring techniques. Also, we are working on optimizing the system even for a large classroom by further improving the performance of SRNet so that it can super resolve even tiny faces of back-benchers and can work for any angle of coverage.

Acknowledgement. We gratefully acknowledge the support of NVIDIA Corporation with the donation of the Titan Xp GPU used for this research.

References

1. Arsenovic, M., Sladojevic, S., Anderla, A., Stefanovic, D.: Facetime–deep learning based face recognition attendance system. In: IEEE 15th International Symposium on Intelligent Systems and Informatics (SISY) 2017, pp. 000053–000058. IEEE (2017)
2. Barron, J.T.: A more general robust loss function. arXiv preprint arXiv:1701.03077 (2017)
3. Chintalapati, S., Raghunadh, M.: Automated attendance management system based on face recognition algorithms. In: 2013 IEEE International Conference on Computational Intelligence and Computing Research (ICCIC), pp. 1–5. IEEE (2013)
4. Dong, C., Loy, C.C., He, K., Tang, X.: Learning a deep convolutional network for image super-resolution. In: Fleet, D., Pajdla, T., Schiele, B., Tuytelaars, T. (eds.) ECCV 2014. LNCS, vol. 8692, pp. 184–199. Springer, Cham (2014). https://doi.org/10.1007/978-3-319-10593-2_13
5. Dong, C., Loy, C.C., He, K., Tang, X.: Image super-resolution using deep convolutional networks. IEEE Trans. Pattern Anal. Mach. Intell. **38**(2), 295–307 (2016)
6. Fu, R., Wang, D., Li, D., Luo, Z.: University classroom attendance based on deep learning. In: 2017 10th International Conference on Intelligent Computation Technology and Automation (ICICTA), pp. 128–131. IEEE (2017)
7. He, K., Zhang, X., Ren, S., Sun, J.: Deep residual learning for image recognition. In: Proceedings of the IEEE Conference on Computer Vision and Pattern Recognition, pp. 770–778 (2016)
8. Huang, J.B., Singh, A., Ahuja, N.: Single image super-resolution from transformed self-exemplars. In: IEEE Conference on Computer Vision and Pattern Recognition (2015)
9. Kim, J., Kwon Lee, J., Mu Lee, K.: Accurate image super-resolution using very deep convolutional networks. In: Proceedings of the IEEE Conference on Computer Vision and Pattern Recognition, pp. 1646–1654 (2016)
10. Kim, J., Kwon Lee, J., Mu Lee, K.: Deeply-recursive convolutional network for image super-resolution. In: Proceedings of the IEEE Conference on Computer Vision and Pattern Recognition, pp. 1637–1645 (2016)
11. Kingma, D.P., Ba, J.: Adam: A method for stochastic optimization. arXiv preprint arXiv:1412.6980 (2014)
12. Lai, W.S., Huang, J.B., Ahuja, N., Yang, M.H.: Fast and accurate image super-resolution with deep laplacian pyramid networks. arXiv preprint arXiv:1710.01992 (2017)
13. Ledig, C., et al.: Photo-realistic single image super-resolution using a generative adversarial network. arXiv preprint arXiv:1609.04802 (2016)
14. Li, H., Lin, Z., Shen, X., Brandt, J., Hua, G.: A convolutional neural network cascade for face detection. In: Proceedings of the IEEE Conference on Computer Vision and Pattern Recognition, pp. 5325–5334 (2015)
15. Lim, B., Son, S., Kim, H., Nah, S., Lee, K.M.: Enhanced deep residual networks for single image super-resolution. In: The IEEE Conference on Computer Vision and Pattern Recognition (CVPR) Workshops, vol. 1, p. 3 (2017)

16. Long, J., Shelhamer, E., Darrell, T.: Fully convolutional networks for semantic segmentation. In: Proceedings of the IEEE Conference on Computer Vision and Pattern Recognition, pp. 3431–3440 (2015)
17. Odena, A., Dumoulin, V., Olah, C.: Deconvolution and checkerboard artifacts. Distill 1(10), e3 (2016)
18. Pan, S.J., Yang, Q.: A survey on transfer learning. IEEE Trans. Knowl. Data Eng. 22(10), 1345–1359 (2010). https://doi.org/10.1109/TKDE.2009.191
19. Rathod, H., Ware, Y., Sane, S., Raulo, S., Pakhare, V., Rizvi, I.A.: Automated attendance system using machine learning approach. In: 2017 International Conference on Nascent Technologies in Engineering (ICNTE), pp. 1–5. IEEE (2017)
20. Sajjadi, M.S., Schölkopf, B., Hirsch, M.: Enhancenet: single image super-resolution through automated texture synthesis. In: 2017 IEEE International Conference on Computer Vision (ICCV), pp. 4501–4510. IEEE (2017)
21. Schroff, F., Kalenichenko, D., Philbin, J.: Facenet: a unified embedding for face recognition and clustering. In: Proceedings of the IEEE Conference on Computer Vision and Pattern Recognition, pp. 815–823 (2015)
22. Simonyan, K., Zisserman, A.: Very deep convolutional networks for large-scale image recognition. arXiv preprint arXiv:1409.1556 (2014)
23. Sun, Y., Chen, Y., Wang, X., Tang, X.: Deep learning face representation by joint identification-verification. In: Advances in Neural Information Processing Systems, pp. 1988–1996 (2014)
24. Szegedy, C., Ioffe, S., Vanhoucke, V., Alemi, A.A.: Inception-v4, inception-resnet and the impact of residual connections on learning. In: AAAI, vol. 4, p. 12 (2017)
25. Tai, Y., Yang, J., Liu, X.: Image super-resolution via deep recursive residual network. In: The IEEE Conference on Computer Vision and Pattern Recognition (CVPR), vol. 1 (2017)
26. Taigman, Y., Yang, M., Ranzato, M., Wolf, L.: Deepface: closing the gap to human-level performance in face verification. In: Proceedings of the IEEE Conference on Computer Vision and Pattern Recognition, pp. 1701–1708 (2014)
27. Viola, P., Jones, M.J.: Robust real-time face detection. Int. J. Comput. Vis. 57(2), 137–154 (2004)
28. Wen, Y., Zhang, K., Li, Z., Qiao, Y.: A discriminative feature learning approach for deep face recognition. In: Leibe, B., Matas, J., Sebe, N., Welling, M. (eds.) ECCV 2016. LNCS, vol. 9911, pp. 499–515. Springer, Cham (2016). https://doi.org/10.1007/978-3-319-46478-7_31
29. Yang, S., Luo, P., Loy, C.C., Tang, X.: From facial parts responses to face detection: a deep learning approach. In: Proceedings of the IEEE International Conference on Computer Vision, pp. 3676–3684 (2015)
30. Zhang, K., Zhang, Z., Li, Z., Qiao, Y.: Joint face detection and alignment using multitask cascaded convolutional networks. IEEE Signal Process. Lett. 23(10), 1499–1503 (2016)

Robust Detection of Iris Region
Using an Adapted SSD Framework

Saksham Jain[1(✉)] and Indu Sreedevi[2]

[1] Netaji Subhas Institute of Technology, Delhi, India
jain.saksham01@gmail.com
[2] Delhi Technological University, Delhi, India
s.indu@dce.ac.in

Abstract. Accurate detection of the iris is a crucial step in several biometric tasks, such as iris recognition and spoofing detection, among others. In this paper, we consider the detection task to be the delineation of the smallest square bounding box that surrounds the iris region. To overcome the various challenges of the iris detection task, we present an efficient iris detection method that leverages the SSD (Single Shot multibox Detector) model. The architecture of SSD is modified to give a lighter and simpler framework capable of performing fast and accurate detection on the relatively smaller sized iris biometric datasets. Our method is evaluated on 4 datasets taken from different biometric applications and from the literature. It is also compared with baseline methods, such as Daugman's algorithm, HOG+SVM and YOLO. Experimental results show that our modified SSD outperforms these other techniques in terms of speed and accuracy. Moreover, we introduce our own near-infrared image dataset for iris biometric applications, containing a robust range of samples in terms of age, gender, contact lens presence, and lighting conditions. The models are tested on this dataset, and shown to generalise well. We also release this dataset for use by the scientific community.

Keywords: Biometrics · Iris detection · SSD

1 Introduction

Iris recognition plays a major role in modern biometry because the muscular pattern of the iris is unique for all humans, and remains unchanged over time. Detection of the iris region is the first step in iris-based biometric systems, and is important in the performance [6] of the entire pipeline. However, it is still a challenging and time consuming task, with much scope for improvement. As such, we focus only on iris region detection in this paper.

In most iris recognition systems, the subsequent step after iris localisation is normalisation of the isolated iris region, with further processing done on this image [6,11]. Several current methods involve localising the iris of the eye with

© Springer Nature Singapore Pte Ltd. 2019
C. Arora and K. Mitra (Eds.): WCVA 2018, CCIS 1019, pp. 51–64, 2019.
https://doi.org/10.1007/978-981-15-1387-9_5

a circular boundary. These algorithms are time and computationally inefficient. Moreover, many biometric images usually have partial occlusion of the iris region by either eyelids or eyelashes. This leads to a noisy image, as can be seen in Fig. 1, leading to a restriction in the overall performance.

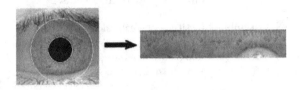

Fig. 1. Noisy normalised image due to partial occlusion by lower eyelid, taken from [11]

However, with the advent of powerful deep learning techniques that have dominated computer vision problems in recent years, CNN-based methodologies can be leveraged to overcome the limitations of the traditional techniques surveyed above. CNNs have shown themselves to be excellent at feature extraction, detection, and recognition [12,14,18], and have already proven their worth in several biometric applications, including iris-based [16].

These techniques perform well under noisy conditions [10,27], and show potential for applications where easier iris region detection might be facilitated by localising the entire iris region, including occlusions and additional noise such as the pupil. Thus, for detection in this paper, we use the smallest square bounding box that completely encapsulates the iris. A novel, lighter and faster framework for detecting the iris region based on the Single Shot Multibox Detector [15] is proposed. It is evaluated and compared with Daugman's [6] algorithm, HOG+linear SVM methodology with sliding window approach [5], and the YOLO [21] network. The proposed framework shows very promising results and speedy, accurate detection despite challenges like noise, occlusion and specular reflections.

The rest of this paper is organised as follows: Sect. 2 presents related work, Sect. 3 summarises the problem, Sect. 4 briefly describes the datasets, Sect. 5 describes the methodology, and Sect. 6 presents the experiments and results. Finally, Sect. 7 presents the concluding remarks of this paper.

2   Related Works

The most well-established technique for iris recognition in use today was given in a seminal paper by Daugman [6]. For iris region detection, it defines an integro-differential operator which fits the circular boundaries of the iris and the pupil by maximising the radial Gaussian via gradient ascent.

This method is modified [28] by applying Hough Transform to a gradient decomposition to approximate the centre of the pupil, while the integro-differential operator fits the iris boundary. In [23] the inner boundary is localised by using the Daugman integro-differential operator, and the outer boundary is modelled using points considered to be the vertices of a triangle inscribed in the circular boundary. This is a faster and computationally cheaper technique than Daugman as it does not involve optimisation.

In [19] the Gabor filter is used to roughly identify the pupil centre, and subsequently, the intero-differential operator localises the iris such that the real centre is in the near vicinity of the rough position of the pupil centre. In, [29], Gabor filters are used for feature extraction and generating a descriptor. Then the proposed probabilistic fuzzy matching scheme is used to compute similarity scores.

In [20], the pupil region is isolated via application of the kNN algorithm on formulated function, and the outer iris region is detected by contrast enhancement and thresholding. In [26], authors present an algorithm which uses the regional properties of the pupil to extract its area and determines the inner iris contour by iterating points, and then comparing and sorting them. Similarly the outer iris contour is determined by an iterative searching methodology, using the pupil centre and approximate radius.

Recently, deep learning based methods have been effectively used in iris recognition systems and related tasks. The authors, in [17], explore the application of pre-trained CNNs to the problem of iris recognition, and demonstrate the effectiveness of their off-the-shelf features for the task. In [2], the authors discuss in detail the network design of a Fully Convolutional Deep Neural Network for iris segmentation, and provide comprehensive comparisons with other methods.

In [1], the authors investigate iris recognition in a visible light environment, and propose a CNN-based method for iris segmentation in the presence of environmental noise of visible light. In [4], a multi-task CNN is proposed to carry out iris localisation, and compute the probability of a presentation attack from the input ocular image. In [24], the authors evaluate baselines for square bounding box location of the iris, and set a benchmark for deep learning-based detectors for the problem.

3   Problem Formulation

Even though Iris based biometric systems are popular, lot of constraints exist. The main challenges faced by current techniques are high computational cost and time consumption for iris detection, inability to deal with scale change, poor performance due to occlusion by eyelashes/eyelid, requirement of iris centring etc. More limitations arise due to ambient conditions such as noise, light reflections etc. In this paper, we propose a modified SSD model uniquely suited for addressing these issues, described in the following sections.

4   Datasets

Four established datasets were chosen from existing biometric applications and literature for this study, namely: Notre Dame Cosmetic Contact Lenses 2013 (NDCLD13) [8], Notre Dame Contact Lens Detection 2015 (NDCLD15) [7], IIIT Delhi Contact Lens Iris (IIITD CLI) [13,30] and CASIA-Iris V3 Interval [3]. An original dataset is also introduced as a part of this study: the IrisDet dataset, which has also been evaluated in this paper.

The NDCLD13 data set contains near-IR images, taken under two sensors. 4200 images (3000 in training set and 1200 in test set) are under the LG4000 sensor and 900 images (600 in training set and 300 in test set) are under the AD100 sensor. The NDCLD15 is an expanded dataset that comprises 7300 images, with 6000 images in the training set and 1300 images in the test set. The IIITD CLI dataset contains a total of 6570 near-IR illumination images taken from 101 subjects using either the Cogent iris sensor or the VistaFA2E sensor. 3000 images comprise the training set (1500 images each corresponding to the two sensors) with the rest comprising the test set for validation and testing. The CASIA-Iris V3 Interval dataset contains 2639 iris images, acquired using a camera that uses circular near-IR LED illumination. All images have a distinctive circular pattern visible in the pupil region. The training set consists of 1500 images, with the remainder being used as the test set. Figure 2 shows sample images.

IrisDet Dataset: This dataset, created during the course of this study, contains 1893 images of the ocular region, taken from 175 subjects, and acquired under near-IR illumination. All images were taken using the IriShield MK2120U single iris camera, and have a resolution of 640×480 pixels (Fig. 3). Although not classified, subjects satisfy either of three conditions: no contact lenses, clear contact lenses, and coloured contact lenses. This dataset differs from all others in this study, in that, about half the images are off-centre, and have been taken in varied lighting conditions, depending on the usage of a goggle. This adds more diversity to the training samples, and helps to train more robust models, as demonstrated in Sect. 5. The training set contains 1300 images, with the remaining 593 being used as the test set. We make this dataset and its annotations available to the scientific community.[1]

5   Proposed Methodology

5.1   Network Framework

The proposed iris detection framework is based on the Single Shot MultiBox Detector(SSD) [15], which can be broken down into two simple major steps: extraction of multi-scale feature maps, and application of small convolution filters for object detection. The starting point is the SSD300 variant, wherein the

---

[1] Print and sign the license agreement available at Saksham Jain's website. Scan and email it to both authors, upon which the download link to the dataset will be sent to the interested researcher.

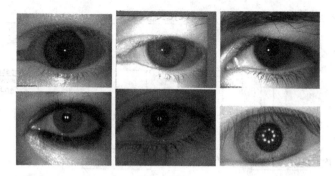

Fig. 2. Dataset sample images (left to right, top to bottom) (i)-NDCLD:AD100, (ii)-NDCLD:LG4000, (iii)-NDCLD15, (iv)-IIITDCLI:Cogent, (v)-IIITDCLI:VistaFA2E, (vi)-Casia-IrisV3

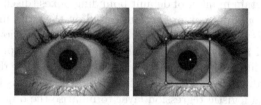

Fig. 3. (i) Shows an image from the introduced IrisDet and (ii) shows manual annotation (https://github.com/tzutalin/labelIm)

image input resolution is 300×300, after which the convolutional layers are applied to the image. The SSD benefits from transfer learning, and uses the VGG16 model [25], trained on ImageNet [14], to do so. The architecture of the network is demonstrated in Fig. 4. The early network layers upto Conv 5_3 form the base of the network, and have the transferred VGG16 pre-learned weights. Transfer learning is used because it enables the model to directly obtain the learned "objectness" [9], from the pre-trained network, and thus allows it to successfully learn the iris features from smaller-sized training sets, despite noise or partial occlusion.

The SSD uses multi-scale feature maps for object detection [15], to better handle variation in location, scale and aspect ratio. Different resolution layers are better at detecting objects at different scales. This eliminates the need for the eye to be at a set distance from the camera, except due to inherent camera constraints. These constraints mean however, that biometric cameras capture images of the entire ocular region, meaning that the iris size itself is constrained. This allows for the removal of lower resolution feature maps which are primarily for detecting large-sized objects. Therefore, in the proposed variant (Fig. 4), only the 38×38, 19×19 and 10×10 feature maps are taken as the prediction source layers. This has the added effect of making the model lighter [31], giving it greater speed without much loss of accuracy for this application. All feature

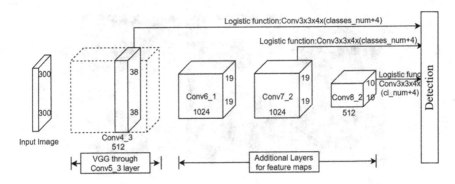

Fig. 4. Architecture of the proposed model

maps contain a certain number of default bounding boxes (discussed in Sect. 5.2) to start off, and bounding box predictions based on them are made.

The objectness scores [9] and the bounding box location offsets (offset between predicted and default boxes) are determined by applying the objectness and location filters respectively to the feature maps. The default boxes are matched to the groundtruth using objectness, and the model is further modified and simplified by a logistic regression layer to binarise the objectness scores.

5.2 Default Bounding Boxes

The HOG+SVM [5] pipeline uses a sliding window strategy for detection, which is limited in terms of speed and is computational cost. Use of region proposals (eg. Faster R-CNN) [22] is much better in terms of both, but prone to mistaking background patches as objects due to inability to contextualise the whole picture. YOLO [21] overcomes this by working in a global context. However YOLO itself has limitations, in that it is spatially constrained on boundary box predictions [21] and somewhat struggles when different scales are involved. SSD, however, overcomes all of these limitations, since it "sees" the whole picture, and adds several feature layers after the base network, after which manually pre-selected bounding boxes are used, as per the requirements of the application.

In the proposed approach, the three feature maps are broken down into a grid formed by 1×1 cells, with the default bounding boxes centred on these cells. There is a single prediction per default box, to keep the number of predictions manageable. These centres are given by [15]:

$$(cx_{def}, cy_{def}) = (\frac{i + 0.5}{|s_n|}, \frac{j + 0.5}{|s_n|})$$

where s_n is the size of the nth feature map, and $i, j \in [0, 1, 2, 3....|s_n|)$. i and j represent the indexes for the default box and matched groundtruth box, respectively. The bounding box location offset, ie. offsets of the predicted bounding boxes to the default boxes for each cell (discussed in Sect. 5.3), is used for box location in place of a global coordinate system.

According to the defined problem, the iris bounding box must be the smallest square bounding the iris. Thus, the aspect ratio of the default bounding boxes is chosen as 1, with the side length of the square determined by the scales, $sc_k \in \{0.1, 0.18, 0.33\}$ selected as per the requirement [15], from the feature map layers. An extra default box with scale, $sc = \sqrt{sc_k \cdot sc_{k+1}}$ is also added for each cell.

5.3   Matching the Default Boxes to the Groundtruth

This step is required so that the groundtruth box can be assigned to a specific default box with which it has the highest IoU (Intersection over Union) value. If the two have a higher IoU value than 0.5 (taken as the threshold in this paper), the default boundary box is considered to be a positive match (the box label is set to 1) otherwise it is a negative match (the box label for is set to 0), due to the logistic decision boundary. Simultaneously, the actual objectness score, and the location offset are also recorded. The objectness score, p always lies between 0 and 1, and since the detection task is solved involving a logistic function layer, the bounding box label x is either 0 or 1. The bounding box location offset is given [15] by:

$$g_j^{cx} = \frac{cx_{gt} - cx_{def}}{a_{def}}$$

$$g_j^{cy} = \frac{cy_{gt} - cy_{def}}{a_{def}}$$

$$a = \log(\frac{a_{gt}}{a_{def}})$$

where (cx, cy) is the matched bounding box centre, j means the same as above, and a is the side of the box. The indexes def and gt respectively denote the default bounding box and groundtruth bounding box.

In case there is conflict where two default boxes are matched with the same groundtruth, the one with the higher IoU value is chosen. Once the positive matches are finalised, the calculated cost function (described in Sect. 5.4) for the corresponding predicted bounding boxes is minimised.

It is natural that far more negative matches are present than positive ones. This can lead to unstable training, due to the resulting class imbalance. Thus, hard negative mining is carried out [15], keeping the ratio of positive matches to negatives at 1:3, for stabler training. This way, the class imbalance can be taken advantage of by having the model learn which predictions are poor. Thus, negative samples during the training phase have a positive impact on actual performance. This is also an added advantage over YOLO, since incorrect localisation is described to be the main source of errors for it [21].

5.4   Loss Function

The loss function or the training objective is a weighted combination of the individual loss functions for the confidence of class prediction, ie. the objectness score, and the bounding box location offsets. The confidence loss can be described by:

$$L_{conf} = -\sum_{i \in Pos}^{N} x_i \log(p_i) - \sum_{j \in Neg}^{N} (1 - x_j) \log(1 - p_j)$$

where, p, x, i, and j hold the same meanings as in the above sections. The location loss can be described by:

$$L_{loc} = \sum_{i \in Pos}^{N} \sum_{m \in (cx, cy, a)} x^m Smooth_{L1}(l_i^m - g_j^m)$$

The location loss is calculated using the Smooth L1 loss, ie. the absolute value loss which is less sensitive to anomalies than L2. Here, l denotes the predicted box offsets, and g denotes the matched groundtruth box parameters. The final loss function [15] is given bye:

$$L_{net} = \frac{1}{N}(\alpha L_{loc} + L_{conf})$$

The value of α is determined via cross validation, and is taken as 1 here.

6   Experiments and Results

In this paper, we evaluate the proposed SSD-based framework, and compare it with the well-established Daugman [6] technique[2], as well as the HOG+SVM and YOLO baselines described in [24]. The implementation of our methodology uses the popular Keras library and is done in python. The experiments are performed on the five datasets mentioned in Sect. 3, on a single Nvidia GeForce GTX 960M GPU accelerated system.

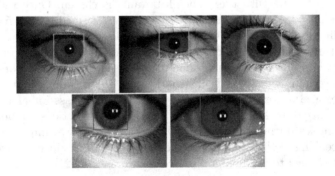

Fig. 5. Positive (green bounding boxes) and Negative (red bonding boxes) Results (Color figure online)

---

[2] Implementation: https://github.com/Qingbao/iris.

The training and testing splits for the datasets are given in Table 1. We use four standard metrics, namely Accuracy, Precision, Recall, and Intersection over Union (IoU), for the evaluation.

Table 1. Train/test splits for the five datasets

| Dataset training/ test split | NDCLD13 | | NDCLD15 | IIIT CLI | | Casia | IrisDet |
	AD100	LG4000		VistaFA2E	Cogent	IrisV3	
Training set	600	3000	6000	1500	1500	1500	1300
Test set	300	1200	1300	1530	2040	1139	593

Since we are not using any large datasets, we utilise data augmentation such as random expansion, flipping and random cropping the images for improved performance. The proposed method makes use of an Non-Maximum Suppression threshold of 0.5, and only those iris bounding squares are accepted, which have a greater than 0.5 objectness score.

Examples of the detection results obtained with our method are shown in Fig. 5. Most failure cases seem to occur in images where the iris is too close to the image border. We perform four experiments, as described below:

Individually Trained and Tested: In this experiment, all models are trained and tested on images corresponding to the same sensor with the Daugman method being applied to the test sets. The proposed approach gave the best results in all the metrics as well as lesser processing time than the other methods (with only YOLO being the faster). The state of the art results obtained are demonstrated in Table 2.

Collectively Trained and Tested: In this experiment, the models are trained on a combined training set taken from all the datasets, and consequently tested on a similarly combined test set, so as to check the ability of the methods to generalise when more varied training samples are provided. Once again, the proposed scheme outstrips the other methods across all metrics (with YOLO being the closest in performance), showing more robustness. Table 3 shows the results obtained, and Fig. 6 portrays the precision vs recall curve for both our proposed SSD variant and YOLO, and highlights the superiority of the proposed scheme.

Trained on Four, Tested on One: In this experiment, the models are trained on a combined training set (containing 11800 images) taken from any four datasets, and subsequently tested on a single test set taken separately from the remaining dataset, one at a time. Due to much greater variation and amount of training samples, all models generalise well, with our proposed method outperforming the others, as shown in Table 4. However, when tested against the IrisDet dataset, the values across all metrics are relatively lower, which may be attributed to the fact that all other datasets have a very low representation of images in which

Table 2. Trained and tested on same sensor

Metric	Method	NDCLD13		NDCLD15	IIIT CLI		Casia	IrisDet
		AD100	LG4000		VistaFA2E	Cogent	IrisV3	
Accuracy	Daugman	94.28	97.53	96.67	95.38	96.34	97.38	94.74
	HOG+SVM	96.57	96.77	96.83	97.93	96.61	92.23	96.16
	YOLO	98.39	98.68	98.48	98.28	98.19	97.21	96.60
	Proposed	99.31	99.41	99.26	99.62	99.33	98.49	98.26
Precision	Daugman	82.49	92.15	89.80	89.34	92.82	96.23	90.06
	HOG+SVM	94.35	92.72	91.18	92.22	87.99	88.48	86.58
	YOLO	95.12	97.83	95.76	93.71	95.88	96.02	92.65
	Proposed	97.47	99.17	97.22	95.23	97.02	98.19	94.60
Recall	Daugman	84.60	93.41	91.63	85.49	86.24	96.38	95.92
	HOG+SVM	92.39	96.72	96.04	94.51	96.44	96.97	95.13
	YOLO	98.78	97.81	97.28	97.85	96.02	97.79	95.67
	Proposed	99.56	98.26	97.99	98.42	97.51	98.49	96.50
IoU	Daugman	80.41	89.67	85.34	80.82	82.61	90.95	90.12
	HOG+SVM	87.52	87.76	86.85	87.23	84.76	86.17	85.16
	YOLO	93.84	95.66	93.25	91.76	91.84	91.24	90.73
	Proposed	94.25	97.21	94.98	93.67	93.10	92.36	92.04

Table 3. Collectively trained and tested

Method	Training set	Test set	Accuracy	Precision	Recall	IoU
Daugman	–	All five	86.54	86.28	94.04	81.09
HOG+SVM	All five	All five	89.67	90.16	92.71	91.14
YOLO	All five	All five	98.32	95.20	97.13	92.54
Proposed	All five	All five	99.27	96.67	98.91	95.52

the iris is significantly off-centre. This lowers the robustness of such a model in a potential single iris biometric application where proper centering may not be a guarantee.

Trained on One, Tested on Four: In this experiment, we see how well the trained models generalise across datasets. This means that the models are trained on separate training sets consisting of 1800 randomly selected images from each of the five datasets (one at a time), and are subsequently tested on a single test set comprising 600 testing samples from the remaining datasets (each dataset having equal representation). For all methods, there is a fall in the values of each metric, which may be attributed to the differences in the way the images were taken, as well as the inherent differences in the camera sensors and environments. The results are presented in Table 5, and show that while all the metric values are higher under our proposed methodology, they are also relatively higher across all metrics when trained on the IrisDet dataset. Thus demonstrating that IrisDet allows for a higher generalisation capability regardless of the model.

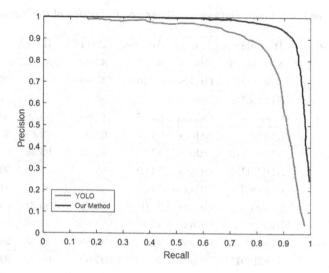

Fig. 6. Precision vs Recall curve comparison for our SSD-based model and YOLO

Table 4. Trained on four datasets and tested on the remaining one

Method	Training set	Test set	Accuracy	Precision	Recall	IoU
HOG+SVM	Others	NDCLD13	92.16	94.26	90.98	92.64
	Others	NDCLD15	92.16	93.12	91.60	94.25
	Others	IIITDCLI	93.45	93.62	94.20	93.56
	Others	CasiaV3	89.96	88.62	89.23	90.13
	Others	IrisDet	94.23	95.32	93.69	94.36
YOLO	Others	NDCLD13	97.68	97.15	98.32	96.98
	Others	NDCLD15	96.85	97.23	96.16	98.32
	Others	IIITDCLI	95.67	95.50	96.68	95.63
	Others	CasiaV3	96.85	98.23	97.96	95.64
	Others	IrisDet	94.13	94.51	93.65	95.67
Proposed	Others	NDCLD13	99.76	98.34	98.68	97.13
	Others	NDCLD15	99.27	98.58	97.24	96.86
	Others	IIITDCLI	99.38	98.95	97.78	96.67
	Others	CasiaV3	98.98	98.85	97.45	97.17
	Others	IrisDet	99.03	98.15	96.89	96.06

For all four metrics, our SSD-based method yields the best results. The processing speed of the proposed method (0.11 s per frame) is also much faster than that of Daugman's (5.20 s per frame) and HOG+SVM (6.72 s per frame), although it loses out to the YOLO detector (0.043 s per frame). However, if a more powerful GPU is used, the detection speed can be increased much further.

Table 5. Trained on a Single Dataset and Tested on Collection of the Remaining Four

Method	Training set	Test set	Accuracy	Precision	Recall	IoU
HOG+SVM	NDCLD13	Others	86.65	85.94	87.25	83.16
	NDCLD15	Others	88.16	87.64	89.32	86.00
	IIITDCLI	Others	89.38	90.42	91.03	87.97
	CasiaV3	Others	82.67	82.88	83.26	80.01
	IrisDet	Others	90.67	91.08	89.89	83.16
YOLO	NDCLD13	Others	94.64	94.23	93.46	91.09
	NDCLD15	Others	94.89	95.08	93.61	90.92
	IIITDCLI	Others	93.64	93.41	92.63	91.39
	CasiaV3	Others	91.08	90.86	90.31	89.84
	IrisDet	Others	94.36	94.03	92.86	91.81
Proposed	NDCLD13	Others	95.75	95.86	95.33	93.14
	NDCLD15	Others	96.95	95.83	96.22	93.10
	IIITDCLI	Others	97.08	95.90	96.04	93.63
	CasiaV3	Others	95.66	94.84	94.13	92.86
	IrisDet	Others	97.87	96.69	96.63	94.14

7   Conclusions

In this paper, we adapt the SSD model for the detection of the iris region, which overcomes several limitations of current techniques such as high computational cost for iris location, and inability to deal with noise, specular reflections, change in scale, poor performance due to occlusion by eyelashes/eyelid, etc. The proposed SSD-based model shows state-of-the-art results, and demonstrates its superiority over existing techniques. Additionally, we introduce and evaluate the IrisDet dataset, which provides the most robust training response. The model, trained on IrisDet, is capable of iris detection on varying scales of ocular images (depending on the eye to camera distance), and also handles off-centre irises. The model shows a lot of potential for extending it to end-to end iris recognition.

References

1. Arsalan, M., et al.: Deep learning-based iris segmentation for iris recognition in visible light environment. Symmetry **9**(11), 263 (2017). https://doi.org/10.3390/sym9110263. http://www.mdpi.com/2073-8994/9/11/263
2. Bazrafkan, S., Thavalengal, S., Corcoran, P.: An end to end deep neural network for iris segmentation in unconstrained scenarios. Neural Netw. **106**, 79–95 (2018)
3. CBSR: Casia-irisv3 image database. http://biometrics.idealtest.org
4. Chen, C., Ross, A.: A multi-task convolutional neural network for joint iris detection and presentation attack detection. In: 2018 IEEE Winter Applications of Computer Vision Workshops (WACVW), pp. 44–51, March 2018. https://doi.org/10.1109/WACVW.2018.00011

5. Dalal, N., Triggs, B.: Histograms of oriented gradients for human detection. In: 2005 IEEE Computer Society Conference on Computer Vision and Pattern Recognition, CVPR 2005, vol. 1, pp. 886–893, June 2005. https://doi.org/10.1109/CVPR.2005. 177

6. Daugman, J.: How iris recognition works. IEEE Trans. Circuits Syst. Video Technol. **14**(1), 21–30 (2004). https://doi.org/10.1109/TCSVT.2003.818350

7. Doyle, J.S., Bowyer, K.W.: Robust detection of textured contact lenses in iris recognition using BSIF. IEEE Access **3**, 1672–1683 (2015). https://doi.org/10.1109/ACCESS.2015.2477470

8. Doyle, J.S., Bowyer, K.W., Flynn, P.J.: Variation in accuracy of textured contact lens detection based on sensor and lens pattern. In: 2013 IEEE Sixth International Conference on Biometrics: Theory, Applications and Systems (BTAS), pp. 1–7, Sept 2013. https://doi.org/10.1109/BTAS.2013.6712745

9. Erhan, D., Szegedy, C., Toshev, A., Anguelov, D.: Scalable object detection using deep neural networks. In: Proceedings of the 2014 IEEE Conference on Computer Vision and Pattern Recognition. pp. 2155–2162. CVPR '14, IEEE Computer Society, Washington, DC, USA (2014). https://doi.org/10.1109/CVPR.2014.276

10. Fawzi, A., Moosavi-Dezfooli, S., Frossard, P.: The robustness of deep networks: a geometrical perspective. IEEE Signal Process. Mag. **34**(6), 50–62 (2017). https://doi.org/10.1109/MSP.2017.2740965

11. Han, M., Sun, W., Li, M.: Iris recognition based on a novel normalization method and contourlet transform. In: 2009 2nd International Congress on Image and Signal Processing, pp. 1–3, October 2009. https://doi.org/10.1109/CISP.2009.5304768

12. Hinton, G., Vinyals, O., Dean, J.: Distilling the Knowledge in a Neural Network. ArXiv e-prints, March 2015

13. Kohli, N., Yadav, D., Vatsa, M., Singh, R.: Revisiting iris recognition with color cosmetic contact lenses. In: 2013 International Conference on Biometrics (ICB), pp. 1–7, June 2013. https://doi.org/10.1109/ICB.2013.6613021

14. Krizhevsky, A., Sutskever, I., Hinton, G.E.: ImageNet classification with deep convolutional neural networks. In: Proceedings of the 25th International Conference on Neural Information Processing Systems, NIPS 2012, vol. 1., pp. 1097–1105. Curran Associates Inc., USA (2012). http://dl.acm.org/citation.cfm?id=2999134. 2999257

15. Liu, W., et al.: SSD: Single Shot MultiBox Detector. ArXiv e-prints, December 2015

16. Menotti, D., et al.: Deep representations for iris, face, and fingerprint spoofing detection. IEEE Trans. Inf. Forensics Secur. **10**(4), 864–879 (2015). https://doi.org/10.1109/TIFS.2015.2398817

17. Nguyen, K., Fookes, C., Ross, A., Sridharan, S.: Iris recognition with off-the-shelf cnn features: a deep learning perspective. IEEE Access **6**, 18848–18855 (2018). https://doi.org/10.1109/ACCESS.2017.2784352

18. Oquab, M., Bottou, L., Laptev, I., Sivic, J.: Learning and transferring mid-level image representations using convolutional neural networks. In: 2014 IEEE Conference on Computer Vision and Pattern Recognition, pp. 1717–1724, June 2014. https://doi.org/10.1109/CVPR.2014.222

19. Radman, A., Jumari, K., Zainal, N.: Fast and reliable iris segmentation algorithm. IET Image Process. **7**(1), 42–49 (2013). https://doi.org/10.1049/iet-ipr.2012.0452

20. Ramkumar, R.P., Arumugam, S.: A novel iris recognition algorithm. In: 2012 Third International Conference on Computing, Communication and Networking Technologies, ICCCNT 2012, pp. 1–6, July 2012. https://doi.org/10.1109/ICCCNT. 2012.6396075

21. Redmon, J., Divvala, S., Girshick, R., Farhadi, A.: You Only Look Once: Unified. Real-Time Object Detection, ArXiv e-prints, June 2015
22. Ren, S., He, K., Girshick, R., Sun, J.: Faster R-CNN: towards real-time object detection with region proposal networks. IEEE Trans. Pattern Anal. Mach. Intell. **39**(6), 1137–1149 (2017). https://doi.org/10.1109/TPAMI.2016.2577031
23. Rodríguez, J.L.G., Rubio, Y.D.: A new method for iris pupil contour delimitation and its application in iris texture parameter estimation. In: Sanfeliu, A., Cortés, M.L. (eds.) Progress in Pattern Recognition, Image Analysis and Applications, pp. 631–641. Springer, Heidelberg (2005). https://doi.org/10.1007/978-3-540-76725-1
24. Severo, E., et al.: A Benchmark for Iris Location and a Deep Learning Detector Evaluation. ArXiv e-prints, March 2018
25. Simonyan, K., Zisserman, A.: Very Deep Convolutional Networks for Large-Scale Image Recognition. ArXiv e-prints, September 2014
26. Su, L., Wu, J., Li, Q., Liu, Z.: Iris location based on regional property and iterative searching. In: 2017 IEEE International Conference on Mechatronics and Automation (ICMA), pp. 1064–1068, August 2017. https://doi.org/10.1109/ICMA.2017.8015964
27. Tang, Y., Eliasmith, C.: Deep networks for robust visual recognition. In: ICML (2010)
28. Tisse, C.L., Martin, L., Torres, L., Robert, M.: Person identification technique using human iris recognition. In: Proceedings of Vision Interface, pp. 294–299 (2002)
29. Tsai, C., Lin, H., Taur, J., Tao, C.: Iris recognition using possibilistic fuzzy matching on local features. IEEE Trans. Syst. Man Cybern. Part B (Cybern.) **42**(1), 150–162 (2012). https://doi.org/10.1109/TSMCB.2011.2163817
30. Yadav, D., Kohli, N., Doyle, J.S., Singh, R., Vatsa, M., Bowyer, K.W.: Unraveling the effect of textured contact lenses on iris recognition. IEEE Trans. Inf. Forensics Secur. **9**(5), 851–862 (2014). https://doi.org/10.1109/TIFS.2014.2313025
31. Yi, J., Wu, P., Hoeppner, D.J., Metaxas, D.: Fast neural cell detection using lightweight SSD neural network. In: 2017 IEEE Conference on Computer Vision and Pattern Recognition Workshops (CVPRW), pp. 860–864, July 2017. https://doi.org/10.1109/CVPRW.2017.119

Dynamic Image Networks for Human Fall Detection in 360-degree Videos

Sumeet Saurav[1][✉], T. N. D. Madhu Kiran[2][✉], B. Sravan Kumar Reddy[2][✉],
K. Sanjay Srivastav[2][✉], Sanjay Singh[1][✉], and Ravi Saini[1]

[1] CSIR-CEERI, Pilani, India
{sumeet,sanjay}@ceeri.res.in
[2] Birla Institute of Technology and Sciences, Pilani, India
{f2015111,f2015072,f2015102}@pilani.bits-pilani.ac.in

Abstract. Detection of falls of elderly people is a trivial yet an immediate problem due to the growing age of the population. This demands the need for autonomous self care systems for providing a quick assistance. The three basic approaches used for fall detection include non-invasive vision based devices, ambient based devices and wearable devices. The paper tries to improve upon the state-of-art of accuracy to 98% using vision based system. This was achieved through transfer learning by extending the idea of action recognition using dynamic images which is a standard RGB image containing the appearance and dynamics of a whole video sequence. Such information is vital in dealing with applications like human action recognition. Since we are also looking for a cheap and scalable solution, the use of a 360° camera seems reasonable and reliable. The top view provided by this camera gives a better perspective than any other alternatives by giving an un-obstructive view of the subjects.

Keywords: Fall detection · Dynamic images · Convolutional Neural Networks

1 Introduction

Physical weakness is a major concern for elderly leading to high frequency of falls impacting their health. According to Ambrose et al. [1], falls are one of the major causes of mortality in old adults. One out of three adults aging 65 or above experience high incidence of falls. The impact of these falls is a major concern for health care systems. These falls might not only lead to severe injuries but also might disturb the mental health such as fear of falling, loss of independence etc. Moreover, the costs associated with it are not negligible: countries like the United States and the United Kingdom, with very different health care systems, spent US$23.3 and US$1.6 billion, respectively, in 2008 [7]. Taking into account the growth of aging population, these expenditures are expected to approach US$55 billion by 2020.

© Springer Nature Singapore Pte Ltd. 2019
C. Arora and K. Mitra (Eds.): WCVA 2018, CCIS 1019, pp. 65–78, 2019.
https://doi.org/10.1007/978-981-15-1387-9_6

Computer Vision techniques have gained an edge over the other techniques due to the advances in deep learning techniques and availability of data in abundance. Classification of daily activities using these concepts has become an easier task. Human Fall Detection is one such application which has great potential in helping elderly people and people with special needs.

The growth of deep learning has had a major impact on computer vision thereby improving the results of many tasks, such as segmentation and object recognition [14]. In this paper, we present a novel approach which takes advantage of Convolutional Neural Networks (CNN) for fall detection (Sect. 3). More precisely, we introduce a CNN that learns how to detect falls from dynamic images (a concept introduced in [5]). Fall datasets are typically of small size. Therefore we take advantage of CNNs by sequentially training the model on Imagenet dataset [8] followed by UCF101 action dataset [23], following the approach of [25]. This is then fine-tuned on our custom dataset for the two class problem of fall detection applying transfer learning.

The rest of the paper is organized as follows. Section 2 describes the related work in the domain of fall detection. Section 3 hovers over the concept of dynamic images, explains the neural network architecture used and the usage of transfer leaning for fine-tuning as done in the paper. Section 4 explains the experiments and results performed and Sect. 5 gives the conclusion of the paper.

2   Related Work

The literature of fall detection is divided between sensor-based and vision-based approaches. The sensor-based fall detection trivially use accelerometers. These devices provide acceleration measures such as vertical acceleration in the case of falls, which are very different compared to daily activities, allowing us to distinguish them. Vallejo et al. [24] and Sengto and Leauhatong [22] proposed feeding a Multilayer Perceptron (MLP), the data of a 3-axis accelerometer (acceleration values in x, y, and z-axis). Kwolek and Kepski [13] used a Kinect camera to obtain an Inertial Measurement Unit (IMU) combined with the depth maps. They also used a Support Vector Machine (SVM) classifier, feeding it the Kinect and the data from the IMU. Approaches like the latter and [12] combined sensors with vision techniques. However, they used vision-based solutions only to verify the prediction of the sensor-based approach.

The purely vision-based approaches focus on the frames of videos to detect falls. Computer vision techniques extract meaningful features such as silhouettes or bounding boxes from the frames to facilitate detection. Some solutions use those features as input for a classifier (e.g., Gaussian Mixture Model (GMM), SVM, and MLP) to automatically detect a fall. Tracking systems are also used extensively; for example, Lee and Mihailidis [15] applied tracking techniques in a close environment to detect falls. They used connected-components labeling to compute the silhouette of a person and extracting features such as the spatial orientation of the center of the silhouette or its geometric orientation. This information helps them to detect positions and also falls. Mubashir et al.

[19] tracked the person's head to improve their base results using a multiframe Gaussian classifier, which was fed with the direction of the principal component and the variance ratio of the silhouette. Rougier et al. [21] suggested using silhouettes as well, which is a common strategy in the literature. Applying a matching system along the video to track the deformation of the silhouette, they analyzed the shape of the body and finally obtained a result with a GMM. Another common technique is computing the bounding boxes of objects to identify a person within it and then detect for a fall using the extracted features (see, for instance, [16,18]). Many solutions are based on supervised learning, that is, extracting lots of features from raw images and using a classifier to learn a decision from labeled data. Supervised learning is the most sought approach for extracting features from raw images and using the labeled data. This is the case, for example, of Charfi et al. [6], who extracted 14 features, applied some transformations (the first and second derivatives, the Fourier transform, and the Wavelet transform), and used an SVM for classification. Harrou et al. [11] used Multivariate Exponentially Weighted Moving Average (MEWMA) charts.

Vision-based fall detection systems can also make use of 3D structures by means of multiple cameras or used powerful depth sensors to extract depth maps. The Kinect camera is very popular given its low price and high performance. Auvinet et al. [2] built a 3D silhouette to analyze the volume distribution along the vertical axis by making use of a kinect camera. Such a camera was also used by Gasparrini et al. [10] to extract 3D features and then use a tracking system to detect the falls. Diraco et al. [9] used depth maps to compute 3D features. Planinc and Kampel [20] used kinect software which provides body joints to obtain the orientation of the major axis on their position. Mastorakis and Makris [17] applied 3D bounding boxes extending the idea of 2D bounding boxes. Other methods include considering videos as 3D volumes instead of 2D frames considering the temporality as the third dimension. The above methods took advantage of the 3D information obtained from the camera systems which may have drawbacks such as multiple cameras with depth sensors might be difficult to deploy. A 2D system is an ideal option when one looks into the deployment perspective as they are cheaper.

The current existing video representations either considers videos as a stack of still images or as a transition between similar frames. Optical flows and motion history images are quite famous when considering such representations. Approaches such as optical flow, where there is an estimation of optical flow between successive frames and summarization of motion between principle components, suffer from the lighting conditions of the environment. Dynamic Images out-perform these methods when such scenarios are considered [5].

Considering CNN for such detection techniques, can be attributed to the fact that CNNs are able to capture short term temporal features which is the general requirement for detecting falls. Eventhough RNNs like LSTM, which store both long and short term memory (patterns) and parse the video frames sequentially to get frame level information, they are computationally intensive and these methods do not suit our task when compared to CNNs. The essence of CNNs was to extract out the features such as local motion patterns, which otherwise can only be extracted using hand-crafted methods. The idea of rank pooling, on which the dynamic images are based, is trivially a hand-crafted technique. Nevertheless, it paves a way for faster and easier processing of data, as multiple frames can be converged into a single dynamic image to process which otherwise would have required every frame to be processed.

3   Materials and Methods

3.1   Dynamic Images

Dynamic Image is a form of 2D representation for a video where multiple frames are overlaid on top of each other to segregate the static background from the foreground motion. This is done through rank pooling where each frame is associated with a weight as explained in [5]. Since data exists in 2D format as a result of dynamic images, we can make use of two dimensional CNN architectures instead of computation heavy three dimensional CNNs. This helps in reducing the computational requirements drastically and can easily be used in real time embedded systems.

The rank pooling performed in order to generate a dynamic image is done through assigning weights to each particular frame of a video. This requires a model to be trained for acquiring the above mentioned weights, which is a tedious and time consuming process. The need for accurate optimization for these weights has the disadvantage of computing the derivative for backpropagation. Particularly in case of CNNs, efficient computation is required along with end-to-end learning for training on large datasets. Hence, we have opted for approximate rank pooling, where these weights are approximated through a mathematical expression generalizing gradient based optimization. This way, the computation required for generating a dynamic image was drastically improved, with no loss in the performance. The derivation of the approximate rank pooling can be observed as below (Fig. 1):

Let the frames of the video be represented as $f_1, f_2, ..., f_T$. Let the feature vector extracted from each frame in the video be $\varphi(f_t)$. Let $A_t = 1/t \sum_{\tau=1}^{T} \varphi(f_\tau)$ be the time average of these feature vectors of all the frames in the video. A ranking function associates to each time t a weight $w(t|v) = \langle v, A_t \rangle$, v is the vector of parameters. Learning v is a convex optimization problem formulated as follows:

$$v^* = \rho(f_1, ..., f_T; \varphi) = argmin_v E(v),$$

$$E(v) = \frac{\lambda}{2}||v||^2 + \frac{2}{T(T-1)} \sum_{q>t} max(0, 1 - w(q|v) + w(t|v)). \qquad (1)$$

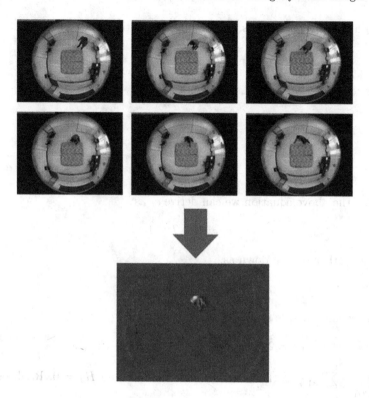

Fig. 1. Dynamic image obtained from a set of frames

The approximate rank pooling derivation is done by considering the first step of gradient optimization of the above Eq. (1), starting with $v = \overrightarrow{0}$, the first approximated solution obtained by gradient descent $v^* = \overrightarrow{0} - \eta \nabla E(v)|_{v=\overrightarrow{0}} \propto -\nabla E(v)|_{v=\overrightarrow{0}}$, for any $\eta > 0$, where

$$\nabla E(\overrightarrow{0}) \propto \sum_{q>t} \nabla max(0, 1 - w(q|v) + w(t|v))|_{v=\overrightarrow{0}}$$

$$= \sum_{q>t} \nabla \langle v, A_t - A_q \rangle = \sum_{q>t} A_t - A_q.$$

v^* can be further expanded as

$$v^* \propto \sum_{q>t} A_q - A_t = \sum_{t=1}^{T} c_t A_t$$

where c_t are scalar coefficients. Expanding summation leads to

$$\sum_{q>t} A_q - A_t = (A_2 - A_1)$$
$$+ (A_3 - A2) + (A_3 - A_1)$$

$$.$$
$$.$$
$$.$$

$$+ (A_T - A1) + (A_T - A2) + ... + (A_T - A_{T-1})$$

Observing the above equation we can derive c_t as

$$c_t = 2t - T - 1 \tag{2}$$

v^* can be further broken down as,

$$v^* \propto c_t A_t = \alpha_t \varphi(f_t)$$

where α_t coefficients are given by

$$\alpha_t = 2(T - t + 1) - (T + 1)(H_T - H_{t-1}), \tag{3}$$

where $H_t = \sum_{i=1}^{t} \frac{1}{i}$ is the t-th Harmonic number and $H_0 = 0$. Rank pooling reduces to

$$\hat{\rho}(f_1, f_2, .., f_T; \varphi) = \sum_{t=1}^{T} \alpha_t \varphi(f_t). \tag{4}$$

So the dynamic image computation reduces to multiplying each of the frame of the video extracted an multiplying them with α_t. We can approximate α_t to a linear in t as follows ignoring the harmonic number part.

$$\alpha_t = 2t - T - 1$$

3.2 Neural Network Architecture

Resnet50 model which was trained on Imagenet dataset was taken as the base model for training on UCF action recognition dataset, where few layers like temporal pooling and app-rank pooling were added to incorporate the concept of dynamic images. The obtained model is an 101-class action recognition model. This model is used as our base model for training on the fall dataset which is a binary classification task.

Two different approaches were taken for designing the network architecture namely Single Dynamic Image (SI) and Multi Dynamic Image (MDI). In case of SI, the entire video is pooled into a single dynamic image which is then sent into a series of convolution layers before sending into the fully connected layers. They are then passed through a classifier layer. In case of MDI, a video is broken down into a small set of frames where each set is rank pooled for a dynamic image. Each dynamic image is sent through a series of convolution layers after which they are temporally pooled. This is then followed by a further set of convolutions and a classifier layer. These are explained in the Figs. 2 and 3.

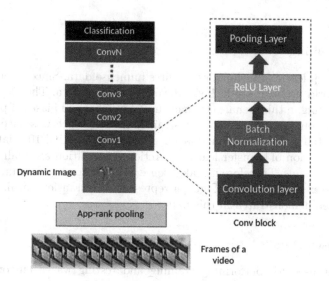

Fig. 2. Single dynamic image

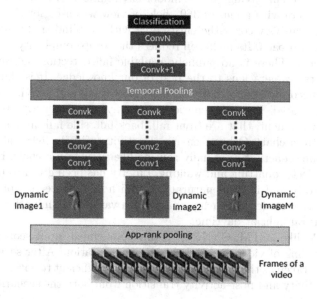

Fig. 3. Multi dynamic image

The convolution block as can be seen in the Figs. 2 and 3, typically consists of a convolution layer, a BatchNorm layer, a max-pool layer for pooling, and a ReLU activation function layer. These layers extract out various features required for proper training which is the general case when convolutional neural networks are considered. Also a drop out layer with a drop-out value of 0.5 was used while training for prevention of over-fitting. Adam Optimizer was used as the optimizer.

3.3 Finetuning

Typically deep learning techniques requires immense data. Since we are dealing with real world problem, we lack significant amount of data. Therefore we used transfer learning in this scenario to make use of weights and biases of pre-trained model for action recognition. This pre-trained model, which was output of fine-tuning resnet50 (pre-trained on Imagenet database) on UCF101 dataset, was also an application of transfer learning. Action Recognition and Fall Detection are related domains. Therefore it makes sense to apply transfer learning as it will converge faster while training. Figure 4 represents the transfer learning pipeline that has been demonstrated in this paper.

3.4 Dataset Details

The dataset used for performing training and testing was a custom dataset named **Fall360 dataset**. This dataset is introduced for two scenarios of "fall" and "activities of daily living" in the indoor environment. A roof-mounted omni-directional camera with a lens of 360° field of view was used for the recordings. It can comprehensively cover the entire area with no blind spots that makes it a dependable approach. Its inclusion reduces the setup complexity, maintenance cost and latency. There is no publicly available fall detection dataset recorded by a 360-degree camera lens to the best of our knowledge. It is the first of its kind. A Balanced dataset for both the scenarios of fall and activity of daily living (ADL) is created. The fall scenario includes five most occurring fall types for an indoor environment, that are front fall, back fall, side fall, imbalanc and fall on standing from chair. Other scenario of activity of daily living also contains five most common activities of daily routine like sitting on a chair, lying down, picking an object, squatting and walking. Liberty has been given to the individuals to act freely and in their natural way. With 22 volunteers, approximately 1300 videos are recorded for each scenario with variable illumination conditions including natural light at daytime.

Algorithms like LSTM, CNN etc. require predefined input sequence, therefore, sub-sampling of videos is needed to a fixed duration. After several experiments, 5 s was taken as the optimal duration that is sufficient to cover pre-activity transition, activity and post-activity transition from both the scenarios. Further

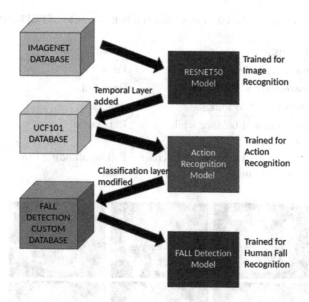

Fig. 4. Transfer learning pipeline

processing of the video clips has been done by estimating motion descriptors like Lucas-Kanade optical flow, Horn-Schunck optical flow, Farneback optical flow and frame differencing. To extend the dataset activities performed in low light conditions captured by the camera's IR night vision were included. The simulated environment was made more realistic by adding more background objects in the room.

The entire dataset was divided into a 75–25% ratio for train split and test split. To account for more reliability three splits were made with test splits having no common data points. The data set consists of videos of 22 different persons each having around 60 videos for both fall and non fall. First 25% of each person's videos of both fall and non-fall were considered for the test-set and the remaining for the train-set for the case of split-1. i.e., for the case of fall, if person-1 has 60 videos in split-1, first 15 videos of him falling will be in the test-set and the remaining 45 in the train-set. For the case of split-2, his 16th to 30th i.e., the next 25%, were considered for the test-set and remaining for the train-set. Similarly the next 25%, i.e., 31st to 45th in test-set were considered for the subsequent split-3.

The camera that was used for this dataset preparation was "D3D FishEye Smart Camera, 360° Panoramic Camera". This camera has a better shell life compared to any other camera. Tables 1 and 2 describe the construction of the data-set using this camera. Figure 5 show cases some of the frames taken from one of the scenario SF (sideways falling action) of one of the 22 persons.

Table 1. Types of the fall, taken into the Fall Scenario of Fall360 dataset

S. no	Fall description	Name of fall	Fall code
1	Fall anteriorly while walking	Front fall	FF
2	Fall posteriorly while walking	Back fall	BF
3	Fall sideways while walking	Side fall	SF
4	Fall due to loss of balance while walking	Imbalance	IF
5	Fall while trying to get up from chair or after getting up from chair	Fall on standing from chair	CF

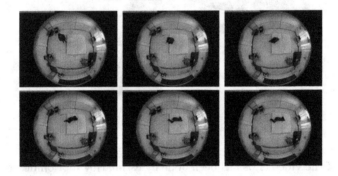

Fig. 5. Figure explaining the scenario of SF mentioned in table1

Table 2. Types of the ADL, taken into the Nonfall Scenario of Fall360 dataset

S. no	Fall description	Name of fall	Fall code
1	Walking towards an office chair and sitting on it	Sitting on chair	SC
2	Lying down on floor	Lying down	LD
3	Walking towards the object and picking it up	Picking an object	PO
4	Sitting on floor cross legged/straight legged/on knees	Squatting	SQ
5	Walk with different pace	Walking	WK

3.5   Implementation Details

We used MATLAB-2017a as the platform, using NVidia GTX-1080 GPU. The environment utilized had CUDA-8 library for using the GPU with gcc-4.9 version. We used MatConvNet, library for implementing the base CNN architecture, and modified the classification layer and the base model considered for fine-tuning. The code which we followed for the sake of training can be found at [3]. We have changed the data-set which the training model takes to our Fall360 dataset along with the changes mentioned above. The base models which we used for training can be found at [4]. Although we tried with multiple base models, best accuracy was achieved considering the model, resnext50-rgb-arpool-split1 at

[4] which is the resnet50 model trained for action recognition (also convergence was fast for this model).

4   Experiments and Results

The dataset was made into three different splits each for training and testing to match the distribution followed by UCF101 dataset for action recognition. The splits made were ensured of having no common data points between the training and testing lists. Two different approaches, SI and MDI were used in the paper as mentioned in Sect. 3.1. Therefore six different experiments for training on fall detection dataset were performed and results of testing can viewed in tables, Tables 3 and 4.

As seen in the two tables, Tables 3 and 4, Single Dynamic Image is having more accuracy than the Multi Dynamic image case which can be attributed to the fact that window size in the case of MDI can lead to a skewed distribution of the action over multiple dynamic images. This phenomenon can lead to misclassification of the actions and hence the accuracy. Size of the window considered plays a crucial role in case of MDI. As MDI is more a powerful method, optimal window size is to be considered which should be in proportion to the length of fall.

Table 3. Results for accuracies of single dynamic image

	Fall accuracy	Non-fall accuracy	Mean accuracy
Split-1	98.8	97.1	97.9
Split-2	98.5	98.8	98.7
Split-3	98.8	95.6	97.2

Table 4. Results for accuracies of multi dynamic image

	Fall accuracy	Non-fall accuracy	Mean accuracy
Split-1	97.9	97.4	97.6
Split-2	95.8	98.8	97.3
Split-3	99.1	96.5	97.8

4.1   Comparison with 3DCNN and CNN-LSTM

In another not yet published work, the same dataset was tested for fall detection using 3DCNN and LSTM models. The experiment results showed that the 3DCNN model trained and validated for the same videos provide 95.3% accuracy on the test set. For the LSTM model, 2DCNN was used for feature extraction and LSTM for fall detection. This model also suits well to fall event detection with 91.62% accuracy. Despite both the methods giving accuracies over 90%, our method out-performed these.

4.2   Real Time Analysis

For real time analysis, we considered the camera, Logitech C930e Full HD Webcam, whose frame rate was 30 fps. We considered an interval of 30 s, where 900 frames (30 fps × 30 s) were processed.

The total time taken to classify this 30 s interval was 35.518 s.

Our window size was 30 frames, i.e., there would be an inference for every 30 frames (and hence 30 epochs). For each epoch the time taken turned out to be 1.184 s (35.518/30).

As it takes 1 s for extracting these 30 frames (as camera specification is 30 fps), 0.184 s is the time taken to classify these 30 frames. For this we assumed the multi-dynamic image (dicnn) model.

5   Conclusion

In this paper, we have successfully applied transfer learning techniques from action recognition to fall detection which achieved an accuracy around 98% on our custom fall database, which is better than the current state-of-art accuracy. The main objective of this paper is to create a vision based solution to detect human fall in an constrained environment of a room. The static nature of a room environment boosts dynamic image technique to extract more foreground details compared other forms of feature extraction there by focusing more on the action (fall) achieving better accuracies. To the best of our knowledge, this is the first time the concept of dynamic images was used in the field of fall detection.

The general tendency of vision based fall detection techniques is to take each frame of a video as independent entity and loose the temporality associated with it. Our paper overcomes this limitation by making use of dynamic images as input to the network as mentioned in the Sect. 3.2.

We believe that the above mentioned vision based solution is an easy to deploy and reliable solution which helps in creating better and safer environments for elderly.

References

1. Ambrose, A.F., Paul, G., Hausdorff, J.M.: Risk factors for falls among older adults: a review of the literature. Maturitas **75**(1), 51–61 (2013)
2. Auvinet, E., Multon, F., Saint-Arnaud, A., Rousseau, J., Meunier, J.: Fall detection with multiple cameras: an occlusion-resistant method based on 3-D silhouette vertical distribution. IEEE Trans. Inf. Technol. Biomed. **15**(2), 290–300 (2011)
3. Bilen, H.: Dynamic image nets (2017). https://github.com/hbilen/dynamic-image-nets
4. Bilen, H.: UCF101 trained models (2017). http://groups.inf.ed.ac.uk/hbilen-data/data/resnext50_dicnn.tar
5. Bilen, H., Fernando, B., Gavves, E., Vedaldi, A.: Action recognition with dynamic image networks. IEEE Trans. Pattern Anal. Mach. Intell. **40**(12), 2799–2813 (2018)

6. Charfi, I., Miteran, J., Dubois, J., Atri, M., Tourki, R.: Definition and performance evaluation of a robust SVM based fall detection solution. SITIS **12**, 218–224 (2012)
7. Davis, J., Robertson, M., Ashe, M., Liu-Ambrose, T., Khan, K., Marra, C.: International comparison of cost of falls in older adults living in the community: a systematic review. Osteoporos. Int. **21**(8), 1295–1306 (2010)
8. Deng, J., Dong, W., Socher, R., Li, L.J., Li, K., Fei-Fei, L.: ImageNet: a large-scale hierarchical image database. In: IEEE Conference on Computer Vision and Pattern Recognition, CVPR 2009, pp. 248–255. IEEE (2009)
9. Diraco, G., Leone, A., Siciliano, P.: An active vision system for fall detection and posture recognition in elderly healthcare. In: Proceedings of the Conference on Design, Automation and Test in Europe, pp. 1536–1541. European Design and Automation Association (2010)
10. Gasparrini, S., Cippitelli, E., Spinsante, S., Gambi, E.: A depth-based fall detection system using a kinect® sensor. Sensors **14**(2), 2756–2775 (2014)
11. Harrou, F., Zerrouki, N., Sun, Y., Houacine, A.: A simple strategy for fall events detection. In: 2016 IEEE 14th International Conference on Industrial Informatics (INDIN), pp. 332–336. IEEE (2016)
12. Harrou, F., Zerrouki, N., Sun, Y., Houacine, A.: Statistical control chart and neural network classification for improving human fall detection. In: 2016 8th International Conference on Modelling, Identification and Control (ICMIC), pp. 1060–1064. IEEE (2016)
13. Kwolek, B., Kepski, M.: Human fall detection on embedded platform using depth maps and wireless accelerometer. Comput. Methods Programs Biomed. **117**(3), 489–501 (2014)
14. LeCun, Y., Bengio, Y., Hinton, G.: Deep learning. Nature **521**(7553), 436 (2015)
15. Lee, T., Mihailidis, A.: An intelligent emergency response system: preliminary development and testing of automated fall detection. J. Telemed. Telecare **11**(4), 194–198 (2005)
16. Liu, C.L., Lee, C.H., Lin, P.M.: A fall detection system using k-nearest neighbor classifier. Expert Syst. Appl. **37**(10), 7174–7181 (2010)
17. Mastorakis, G., Makris, D.: Fall detection system using kinect's infrared sensor. J. Real-Time Image Proc. **9**(4), 635–646 (2014)
18. Miaou, S.G., Sung, P.H., Huang, C.Y.: A customized human fall detection system using omni-camera images and personal information. In: 1st Transdisciplinary Conference on Distributed Diagnosis and Home Healthcare, D2H2, pp. 39–42. IEEE (2006)
19. Mubashir, M., Shao, L., Seed, L.: A survey on fall detection: principles and approaches. Neurocomputing **100**, 144–152 (2013)
20. Planinc, R., Kampel, M.: Introducing the use of depth data for fall detection. Pers. Ubiquit. Comput. **17**(6), 1063–1072 (2013)
21. Rougier, C., Meunier, J., St-Arnaud, A., Rousseau, J.: Robust video surveillance for fall detection based on human shape deformation. IEEE Trans. Circuits Syst. Video Technol. **21**(5), 611–622 (2011)
22. Sengto, A., Leauhatong, T.: Human falling detection algorithm using back propagation neural network. In: Biomedical Engineering International Conference (BMEiCON), pp. 1–5. IEEE (2012)

23. Soomro, K., Zamir, A.R., Shah, M.: UCF101: a dataset of 101 human actions classes from videos in the wild. arXiv preprint arXiv:1212.0402 (2012)
24. Vallejo, M., Isaza, C.V., Lopez, J.D.: Artificial neural networks as an alternative to traditional fall detection methods. In: 2013 35th Annual International Conference of the IEEE Engineering in Medicine and Biology Society (EMBC), pp. 1648–1651. IEEE (2013)
25. Wang, L., Xiong, Y., Wang, Z., Qiao, Y.: Towards good practices for very deep two-stream convnets. arXiv preprint arXiv:1507.02159 (2015)

Image Segmentation and Geometric Feature Based Approach for Fast Video Summarization of Surveillance Videos

Raju Dhanakshirur Rohan[✉], Zeba ara Patel, Smita C. Yadavannavar, C. Sujata, and Uma Mudengudi

KLE Technological University, Hubli, India
rdshirur.cstaff@iitd.ac.in, rohanrd28296@gmail.com

Abstract. In this paper, we propose a geometric feature and frame segmentation based approach for video summarization. Video summarization aims to generate a summarized video with all the salient activities of the input video. We propose to retain the salient frames towards generation of video summary. We detect saliency in foreground and background of the image separately. We propose to model the image as MRF (Markov Random Field) and use MAP (Maximum a-posteriori) as final solution to segment the image into foreground and background. The salient frame is defined by the variation in feature descriptors using the geometric features. We propose to combine the probabilities of foreground and background segments being salient using DSCR (Dempster Shafer Combination Rule). We consider the summarized video as a combination of salient frames for a user defined time. We demonstrate the results using several videos in BL-7F dataset and compare the same with state of art techniques using retention ratio and condensation ratio as quality parameters.

Keywords: Video summarization · Graph cut · Geometric features · Dempster Shafer Combination Rule (DSCR)

1 Introduction

In this paper, we propose a feature based approach for video summarization. Video summarization aims to generate a summarized video with all the salient activities of the input video. We propose to retain the salient frames towards generation of video summary. Due to huge content available in the internet in form of videos, searching the most appropriate and effective information is time consuming for the user. Video summarization is the method to generate a short video containing the most effective frames of the available video. Video summarization finds its applications in video surveillance systems [3,24,26] in which computer vision algorithms, such as tracking, behavior analysis, and object segmentation, are integrated in cameras and/or servers. It also finds its applications in movie trailer generation, sport summary generation etc.

© Springer Nature Singapore Pte Ltd. 2019
C. Arora and K. Mitra (Eds.): WCVA 2018, CCIS 1019, pp. 79–88, 2019.
https://doi.org/10.1007/978-981-15-1387-9_7

Many researchers have worked on video summarization. Objects and people within a video play vital role for video summarization [18]. This is because, we generally represent the events in a video by people/objects and their activities. Moreover, people/objects in the video have the high-level of the semantic perception. Also, along with this, humans usually are more attentive towards the moving objects in a video [6,8]. However, researchers consider the problem of extracting the moving objects from a video that has changes in illumination, high noise, bad contrast and multimodal environment as a challenging problem [2,22]. However, in the videos with low contrast, the edges of objects are given higher prominence [22]. Also, this method is further sensitive to the variation in the shape and position of the object.

To resolve these problems, we can apply the theory of edge-segments (i.e. groups of connected sequential edge pixels) [8]. But, [4,6] claims that the edge-segments based methods fail when the video has shape matching errors or local shape distortion. The *state of the art* methods for object detection use the ellipses or circles to represent curve fragments [6]. Even then, the problem persists if the video has low illumination [6]. Also, we observe that in real world, an object can take up any shapes other than circular, elliptical, parabolic, or hyperbolic curve. Thus, the object detection methods that approximate the shapes to the primitive structures fail in such circumstances. However, we can easily fit a conic part for simple objects.

In [10], authors use a set of similar objects to build a model for summarization. Authors in [16] present a part-based object movement framework. Authors in [14] apply object bank and object-like windows to extract the objects and then they perform story based video summarization. Authors in [5] propose a complementary background model. Pixel-based motion energy and edge features are combined in [23] for summarization. Authors in [12] propose a background subtraction method to detect foreground objects for video summarization. Authors in [13] modify the previous idea for Aggregated Channel Features (ACF) detection and a background subtraction technique for object detection.

In [21], authors propose a video summarization technique by merging three multi-modal human visual sensitive features, namely, motion information, foreground objects, and visual saliency.

Authors in [15] propose a min-cut based approach for generating storyboard. Also authors in [15] modify the previous idea and propose a Bayesian foraging technique for objects and their activities detection to summarize a video. The grid background model is applied in [7]. Authors of [17], deploy a key-point matching technique for video segmentation. Authors in [8] apply Spatio-temporal slices to select the states of the object motion.

Authors in [9] propose a learning based approach for video summarization. They describe the Objects in a video by Histogram of Optical Flow Orientations and then apply a SVM based classifier. Authors in [19] propose unsupervised framework via joint embedding and sparse representative selection for video summarization. The objective function is two-stream in nature. The first objective is to capture multi-view correlations using an embedding, that assists in extracting a diverse set of representatives and the second is to use $L-2$ norm to model the sparsity while selecting representative shots for the summary. Authors in [28] uses RNN to exploit the temporal relationship between frames for saliency detection. Authors in [20] makes use of fully connected neural network for video summarization. However all these techniques need high computational capability which makes it highly impossible for low-cost real time implementation.

Authors in [27] apply a modularity cut algorithm to track objects for summary generation. Gaussian Mixture model based approach is employed in [4]. The key frames are selected based on the parameters of cluster. Authors in [4,6], use geometric primitives (such as lines, arcs) for distinguishable descriptors than edge-pixels or edge-segments.

These primitives are independent of the size of the object, and also they are efficient for matching and comparisons. They are also invariant to scale and viewpoint changes. Thus, these geometric primitives represent objects with complex shapes and structures effectively. Also, they are useful in cognitive system [11].

In this paper, we propose to fuse the techniques of foreground/background segmentation and the use of geometric features for saliency detection in order to achieve video summarization. Towards this, we make the following contributions:

- We propose to detect the saliency of a frame by detecting the saliency of its foreground and background separately and then combine the probabilities of foreground and background being salient to check the saliency of a frame.
  - We propose to model the image as an MRF and use MAP using graph-cut as final solution for foreground and background segmentation.
  - We propose to combine the probabilities of foreground and background being salient using the Dempster Shafer Combination rule (DSCR).
- We propose to use the changes in the variant of the geometric features (such as lines, arcs) to decide the saliency of a frame. For efficient extraction of geometric primitives,
  - We propose to extract the PCA features to detect the principle components of foreground and background frames.
  - We convert the image from RGB to YCbCr and compute PCA on Y channel of the frame to retain the chromic information.
- We demonstrate the results using the BL-7F dataset and compare the results using the state-of-the-art techniques with the help of the quantitative parameters such as condensation ratio and retention ratio.

2   Proposed Framework

We demonstrate the proposed framework in Fig. 1. We propose to detect the saliency of a frame by detecting the saliency of its foreground and background separately. We propose to detect the changes in the PCA and Geometric Primitives such as lines and contours by computing difference in standard deviation of the segments and comparing the difference with a heuristically set threshold.

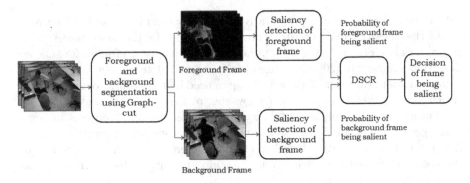

Fig. 1. Proposed framework

The threshold for background is kept much lower as compared to that of foreground with an intuition that any small motion in background is much significant as compared to small motion in foreground. We find separate probabilities for foreground and background segments being salient. We combine the two probabilities using DSCR to obtain joint probability. We decide if the given frame is salient based on the decision boundary set upon the joint probability.

2.1 Foreground and Background Segmantation

We propose to separate the foreground of the scene from the background using Energy Minimization via Graphcut. We model every frame as MRF (Markov Random Field) and use MAP (Maxima A Posteriori) estimate as the final solution. In this framework, we use the grid graph containing image pixels for MRF. Here, we try to find the labelling for the pixels in the image f with minimum energy.

$$E(f) = Esmooth(f) + Edata(f)$$

Where $Edata(f)$ is defined by,

$$Edata(f) = \sum_{p \in P} Dp(fp)$$

Here $Esmooth(f)$ measures the extent to which f is not piecewise smooth, whereas the $Edata(f)$ measures the total disagreement between f and the observed data. Researchers have proposed many different energy functions. The form of $Esmooth(f)$ is typically,

$$Esmooth(f) = \sum_{p,q \in N} u\{p,q\}.T(f_p \neq f_q)$$

here, T is indicator function. It will output 1 if the input condition is true. We use Potts Model in which, discontinuities between any pair of labels are penalized equally. This is, in some sense, the simplest discontinuity preserving model.

We then obtain the two segments of the image, one corresponding to foreground and the other corresponding to background. The foreground and background segmentation for two datasets is shown in Fig. 2.

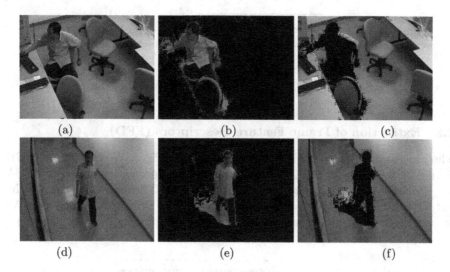

(a)   (b)   (c)

(d)   (e)   (f)

Fig. 2. Segmentation of image into foreground and background frames: (a), (d) are original images. (b), (e) are the corresponding foreground frames. (c), (f) are the corresponding background frames

2.2   Saliency Detection of Foreground and Background Frames

We demonstrate the saliency detection block in Fig. 3. The input for the saliency detection is the segmented frame (Foreground or background). We propose to use the changes in the variant of geometric primitives to decide the saliency of a frame. We extract the variant of geometric features, named the frame feature descriptors (FFD). The process of FFD extraction is demonstrated in Fig. 4. We then find the standard deviation between the extracted feature vectors of the consecutive frames. The probability of frame being salient is decided by the extent with which the obtained standard deviation is greater than a heuristically set threshold.

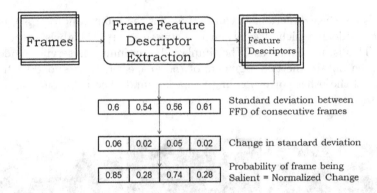

| 0.6 | 0.54 | 0.56 | 0.61 | Standard deviation between FFD of consecutive frames |

| 0.06 | 0.02 | 0.05 | 0.02 | Change in standard deviation |

| 0.85 | 0.28 | 0.74 | 0.28 | Probability of frame being Salient = Normalized Change |

Fig. 3. Saliency detection of foreground and background frames

2.3   Extraction of Frame Feature Descriptors (FFD)

The process of FFD extraction is demonstrated in Fig. 4. We convert the RGB frames of the video to YCbCr to retain the colour information. We apply PCA on 'Y' channel of the image to get PCA transformed 'Y' channel. We convert the output to RGB to obtain the images with enhanced principal components. We extract geometric features from images with enhanced principal components.

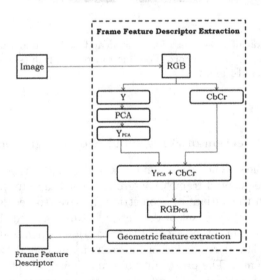

Fig. 4. Extraction of frame feature descriptors (FFD)

We extract the objects present in the salient frames as geometric features. We apply Canny edge detection to find the edges of the objects. Using these edges we find the geometric features like line segments and contours. Contours represent the largest area in the frames. Hence we find the difference in the frames by monitoring the change in the position of the line segments present in the area of the contours.

2.4 Joint Probability Using DSCR

We combine the two probabilities using Dempster Shafer Combination Rule (DSCR) to obtain the joint probability. We decide if the given frame is salient based on the decision boundary set upon the joint probability. Let P_1 and P_2 be the probabilities to be combined. DSCR combines two hypothesis consisting of three parameters, mass of belief, mass of disbelief and mass of uncertainty rather than two probabilities. We construct hypothesis, hyp_1 and $hyp2$ as a set of mass of belief $(m(b))$, disbelief $(m(d))$ and ambiguity $(m(u))$ respectively. We set mass of belief $(m_1(b))$ for hyp_1 as P_1 and mass of belief $(m_2(b))$ for hyp_2 be P_2. We assume mass of disbelief $(m_1(d))$ for hyp_1 and hyp_2 to be 0 and mass of ambiguity $(m_1(u)$ and $m_2(u))$ for hyp_1 and hyp_2 as $1-P_1$ and $1-P_2$ respectively. We combine hyp_1 and hyp_2 using combination table as shown in Table 1.

Table 1. Combination table

$\cap$	m_1^{belief}	$m_1^{disbelief}$	$m_1^{ambiguity}$
m_2^{belief}	ψ_1	$\emptyset$	ψ_1
$m_2^{disbelief}$	$\emptyset$	ψ_2	ψ_1
$m_2^{ambiguity}$	ψ_2	ψ_2	Ω

In the combination table, the product of mass of belief of one hypothesis and mass of disbelief of other hypothesis gives rise to conflict and is represented by $\emptyset$. The product of mass of belief and mass of belief or the product of mass of belief and mass of uncertainty represents joint belief and is represented by ψ_1. Similarly ψ_2 represents the joint disbelief.

The Combined belief of two evidences is considered as Joint probabilities and is given by:

$$JointProbability = \frac{\sum \psi_1}{1 - \sum \emptyset}$$

We decide if the given frame is salient based on the decision boundary set upon the joint probability. The advantage of using DSCR for combining the two probabilities is that it emphasis of the fact that if P_1 is the probability of frame being salient, then $1 - P_1$ need not be the probability of frame being non-salient. It can be uncertainty as well.

3   Results and Discussions

We evaluate our approach using BL-7F dataset. In this dataset, 19 surveillance videos are taken from fixed surveillance cameras located in the seventh floor of the BarryLam Building in the National Taiwan University. Each video consists of 12,900 frames with a duration of 7 min and 10 s. We compare our results using Retention ratio and Condensation ratio as evaluation metrics.

Table 2. Comparison of condensation ratio (in percentage) of the proposed method with the different state-of the art techniques [1, 18, 25] for different surveillance videos. Here RR = retention ratio and is seen to be 1 for the results, unless mentioned.

Video	Duration of given video (min:sec)	Duration of summarized video (min:sec)	Valdes et al. IAMIS 2008	Almedia et al. ISM 2010	S. Ou et al. JSTSP 2015	Proposed framework
bl-0	07:10	00:03	49.53	51.60	93.02	99.29
bl-1	07:10	00:08	36.27	91.6	83.02	97.97
bl-2	07:10	00:09	61.8	50	75.34	97.92
bl-3	07:10	00:02	56.744	98.83	96.27	99.45
bl-4	07:10	00:13	64.41	88.37	80.69	97.03
bl-5	07:10	00:04	36.27	90.46	85.58	99.05
bl-6	07:10	00:05	22.32	100 (RR = 0)	95.35	98.8
bl-7	07:10	00:05	30.93	95.34	88.37	98.8
bl-8	07:10	00:01	22.32	99.3	98.37	99.74
bl-9	07:10	00:09	17.9	95.58	90.93	98.01
bl-10	07:10	00:08	93.48	93.48	74.19	99
bl-11	07:10	00:07	68.6	62.09	73.95	98.31
bl-12	07:10	00:04	48.37	50	69.06	96.42
bl-14	07:10	00:14	63.72	94.88	83.25	96.62
bl-15	07:10	00:07	94.65	89.53	84.18	98.31
bl-16	07:10	00:26	89.53	89.53	76.15	93.85
bl-17	07:10	00:28	61.16	51.16	77.67	93.35
bl-18	07:10	00:03	61.62	95.11	85.16	99.24

Retention ratio is the ratio of number of objects in the summarized video to the number of objects in the original video.

$$RR = \frac{number\ of\ objects\ in\ summarized\ video}{number\ of\ objects\ in\ input\ video}$$

Condensation ratio is the ratio of length of summarized video to length of the input video.

$$CR = (1 - \frac{length\ of\ summarized\ video}{length\ of\ input\ video}) * 100$$

We find that the proposed method gives better results as compared to results obtained from the other state-of-the-art techniques. Retention ratio for the proposed method is unity for all videos and Condensation ratios are also very high compared to the existing methods. The comparison of the condensation ratio (in percentage) of the proposed method with the different state-of the art techniques [1, 18, 25] for different surveillance videos is demonstrated in Table 2.

4   Conclusions

In this paper, we have proposed a geometric feature and frame segmentation based approach for video summarization. We detected saliency in foreground and background of the image separately. We proposed to model the image as MRF (Markov Random Field) and use MAP (Maximum a-posteriori) as final solution to segment the image into foreground and background. The salient frame was effectively defined by the variation in feature descriptors using variant of geometric features. We proposed to combine the probabilities of foreground and background segments being salient using DSCR (Dempster Shafer Combination Rule). We modelled the summarized video as a combination of salient frames for a user defined time. We have demonstrated the results using several videos in BL-7F dataset and compared the same with state of art techniques using retention ratio and condensation ratio as quality parameters to prove the superiority of the proposed method over the other algorithms.

References

1. Almeida, J., Torres, R.D.S., Leite, N.J.: Rapid video summarization on compressed video. In: 2010 IEEE International Symposium on Multimedia, pp. 113–120, December 2010
2. Bagheri, S., Zheng, J.Y.: Temporal mapping of surveillance video. In: 2014 22nd International Conference on Pattern Recognition, pp. 4128–4133, August 2014
3. Chan, W.K., Chang, J.Y., Chen, T.W., Tseng, Y.H., Chien, S.Y.: Efficient content analysis engine for visual surveillance network. IEEE Trans. Circuits Syst. Video Technol. **19**(5), 693–703 (2009)
4. Chang, W., Lee, S.Y.: Description of shape patterns using circular arcs for object detection. IET Comput. Vis. **7**(2), 90–104 (2013)
5. Chen, S.C., et al.: Target-driven video summarization in a camera network. In: 2013 IEEE International Conference on Image Processing, pp. 3577–3581, September 2013
6. Chia, A.Y.S., Rajan, D., Leung, M.K., Rahardja, S.: Object recognition by discriminative combinations of line segments, ellipses, and appearance features. IEEE Trans. Pattern Anal. Mach. Intell. **34**(9), 1758–1772 (2012)
7. Cui, Y., Liu, W., Dong, S.: A time-slice optimization based weak feature association algorithm for video condensation. Multimedia Tools Appl. **75**, 17515–17530 (2016)
8. Kovesi, P.D.: MATLAB and Octave functions for computer vision and image processing, January 2000
9. Fan, C.T., Wang, Y.K., Huang, C.R.: Heterogeneous information fusion and visualization for a large-scale intelligent video surveillance system. IEEE Trans. Syst. Man Cybern. Syst. **47**(4), 593–604 (2017)
10. Fei, M., Jiang, W., Mao, W.: Memorable and rich video summarization. J. Vis. Commun. Image Represent. **42**(C), 207–217 (2017)
11. Hu, R.X., Jia, W., Ling, H., Zhao, Y., Gui, J.: Angular pattern and binary angular pattern for shape retrieval. IEEE Trans. Image Process. **23**(3), 1118–1127 (2014)
12. Huang, C.R., Chung, P.C.J., Yang, D.K., Chen, H.C., Huang, G.J.: Maximum a posteriori probability estimation for online surveillance video synopsis. IEEE Trans. Circuits Syst. Video Technol. **24**(8), 1417–1429 (2014)

13. Li, X., Wang, Z., Lu, X.: Surveillance video synopsis via scaling down objects. IEEE Trans. Image Process. **25**(2), 740–755 (2016)

14. Lu, Z., Grauman, K.: Story-driven summarization for egocentric video. In: 2013 IEEE Conference on Computer Vision and Pattern Recognition, pp. 2714–2721, June 2013

15. Napoletano, P., Boccignone, G., Tisato, F.: Attentive monitoring of multiple video streams driven by a Bayesian foraging strategy. IEEE Trans. Image Process. **24**(11), 3266–3281 (2015)

16. Nie, Y., Sun, H., Li, P., Xiao, C., Ma, K.L.: Object movements synopsis viapart assembling and stitching. IEEE Trans. Visual Comput. Graph. **20**(9), 1303–1315 (2014)

17. Otani, M., Nakashima, Y., Sato, T., Yokoya, N.: Textual description-based video summarization for video blogs. In: 2015 IEEE International Conference on Multimedia and Expo (ICME), pp. 1–6, June 2015

18. Ou, S.H., Lee, C.H., Somayazulu, V.S., Chen, Y.K., Chien, S.Y.: On-line multiview video summarization for wireless video sensor network. IEEE J. Sel. Top. Signal Process. **9**(1), 165–179 (2015)

19. Panda, R., Roy-Chowdhury, A.K.: Multi-view surveillance video summarization via joint embedding and sparse optimization. IEEE Trans. Multimedia **19**(9), 2010–2021 (2017)

20. Rochan, M., Ye, L., Wang, Y.: Video summarization using fully convolutional sequence networks. In: Ferrari, V., Hebert, M., Sminchisescu, C., Weiss, Y. (eds.) ECCV 2018. LNCS, vol. 11216, pp. 358–374. Springer, Cham (2018). https://doi.org/10.1007/978-3-030-01258-8_22

21. Salehin, M.M., Paul, M.: Adaptive fusion of human visual sensitive features for surveillance video summarization. J. Opt. Soc. Am. A: Opt. Image Sci. Vis. **34**(5), 814–826 (2017)

22. Salehin, M., Zheng, L., Gao, J.: Conics detection method based on Pascal's theorem. In: Proceedings of the 10th International Conference on Computer Vision Theory and Applications - Volume 1: VISAPP, (VISIGRAPP 2015), pp. 491–497. INSTICC, SciTePress (2015)

23. Shih, H.C.: A novel attention-based key-frame determination method. IEEE Trans. Broadcast. **59**(3), 556–562 (2013)

24. Taj, M., Cavallaro, A.: Distributed and decentralized multicamera tracking. IEEE Signal Process. Mag. **28**(3), 46–58 (2011)

25. Valdés, V., Martínez, J.M.: On-line video summarization based on signature-based junk and redundancy filtering. In: 2008 Ninth International Workshop on Image Analysis for Multimedia Interactive Services, pp. 88–91, May 2008

26. Valera, M., Velastin, S.A.: Intelligent distributed surveillance systems: a review. IEE Proc. - Vis. Image Signal Process. **152**(2), 192–204 (2005)

27. Zhang, S., Roy-Chowdhury, A.K.: Video summarization through change detection in a non-overlapping camera network. In: 2015 IEEE International Conference on Image Processing (ICIP), pp. 3832–3836, September 2015

28. Zhao, B., Li, X., Lu, X.: Hierarchical recurrent neural network for video summarization. In: ACM Multimedia (2017)

Supervised Hashing for Retrieval of Multimodal Biometric Data

T. A. Sumesh$^{(\boxtimes)}$ ⓘ, Vinay Namboodiri$^{(\boxtimes)}$ ⓘ, and Phalguni Gupta$^{(\boxtimes)}$

Department of CSE, Indian Institute of Technology Kanpur, Kanpur, India
{sumeshta,vinaypn,pg}@cse.iitk.ac.in

Abstract. Biometric systems commonly utilize multi-biometric approaches where a person is verified or identified based on multiple biometric traits. However, requiring systems that are deployed usually require verification or identification from a large number of enrolled candidates. These are possible only if there are efficient methods that retrieve relevant candidates in a multi-biometric system. To solve this problem, we analyze the use of hashing techniques that are available for obtaining retrieval. We specifically based on our analysis recommend the use of supervised hashing techniques over deep learned features as a possible common technique to solve this problem. Our investigation includes a comparison of some of the supervised and unsupervised methods viz. Principal Component Analysis (PCA), Locality Sensitive Hashing (LSH), Locality-sensitive binary codes from shift-invariant kernels (SKLSH), Iterative quantization: A procrustean approach to learning binary codes (ITQ), Binary Reconstructive Embedding (BRE) and Minimum loss hashing (MLH) that represent the prevalent classes of such systems and we present our analysis for the following biometric data: Face, Iris, and Fingerprint for a number of standard datasets. The main technical contributions through this work are as follows: (a) Proposing Siamese network based deep learned feature extraction method (b) Analysis of common feature extraction techniques for multiple biometrics as to a reduced feature space representation (c) Advocating the use of supervised hashing for obtaining a compact feature representation across different biometrics traits. (d) Analysis of the performance of deep representations against shallow representations in a practical reduced feature representation framework. Through experimentation with multiple biometrics traits, feature representations, and hashing techniques, we can conclude that current deep learned features when retrieved using supervised hashing can be a standard pipeline adopted for most unimodal and multimodal biometric identification tasks.

Keywords: Biometric systems · Supervised hashing

1 Introduction

There has been tremendous growth in personal digital data stored across the Internet. With the proliferation of social media applications, this trend has

© Springer Nature Singapore Pte Ltd. 2019
C. Arora and K. Mitra (Eds.): WCVA 2018, CCIS 1019, pp. 89–101, 2019.
https://doi.org/10.1007/978-981-15-1387-9_8

increased. These data majorly comprising of images of persons has become a means to identify people. But the size of this data is enormous to the tune of billion in the case of Facebook as it has got more than 2 billion users. Similarly, Twitter, Instagram, and other social media applications have millions of images of the users. Several countries across the world also maintain a unique identification system of their citizens. These systems store various biometric features like face, iris and fingerprint images of the persons in addition to other credentials. In order to use these images for identification purposes, indexing techniques using approaches like multidimensional trees comes into the picture. But indexing has always been a challenging task in the case of biometric databases due to various challenges like high dimensional feature representations, a varying number of dimensions for same trait and scalability. Further, with an extensive collection of data available over the internet, there is a need for faster indexing and search so that finding nearest neighbors can be done quickly.

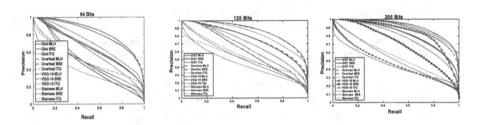

Fig. 1. Precision Recall curves for LFW Face database with Siamese, Gist, Overfeat and VGG-16

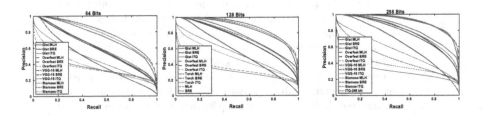

Fig. 2. Precision Recall curves for CASIA Fingerprint database with Siamese, Gist, Overfeat and VGG-16

Various biometrics traits usually need high dimensional feature representation, and they suffer from the curse of dimensionality. For instance, a face has a large number of feature points making it a feature rich biometric trait. For example, a face image of size 100×100 can have feature points up to 10,000. Due to the easy availability of non-intrusive surveillance systems, the face could

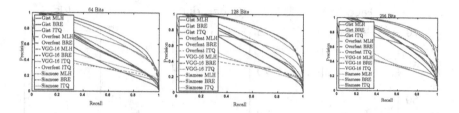

Fig. 3. Precision Recall curves for CASIA Iris database with Siamese, Gist, Overfeat and VGG-16

be easily used to recognize people. However, it requires handling large databases of faces for identification.

Feature representation of biometric data can affect the performance of the indexing mechanism. Previously some methods in the literature have used heuristics based feature representation. The indexing mechanism in such pipelines mostly use tree data structure like Kd-Tree. But these data structures are not very useful in handling the curse of dimensionality and storage requirements as these methods were dealing with feature representation in the real space itself. Thus these methods could not work well with high dimensional data in the order 1K features or more. Fortunately, there have been some attempts recently to use binary hashing techniques in the visual object recognition and scene recognition, as an effort to enhance the speed of the image retrieval and reduce the storage requirement. To the best of our knowledge, there has been no systematic analysis of this approach in the domain of biometric identification. Therefore, in this paper we explore the possibility of including supervised binary hashing in the existing pipeline of the biometric identification system.

There have been a few instances where the use of hashing techniques for biometric data proposed in the past. Tulyakov et al. [14] proposed a hashing method for fingerprint data. In this method, minutiae points are represented as complex numbers and hash functions are constructed based on some complex function which is independent of the order of minutiae points. Sutcu et al. [13] proposed a hash function based on one way transform function, designed as a sum of properly weighted and shifted Gaussian functions for biometrics. Ngo et al. [6] proposed a method for dimensionality reduction using random thresholding projection to improve the accuracy of the face recognition. Rathgeb and Uhl [10] proposed a hashing based on thresholding for Iris based recognition system. But most of these proposals are specific to some specific biometric data, and their main focus was on improving security in the verification pipeline and not the retrieval speed improvement or storage space optimization.

In this paper, we propose a feature extraction mechanism based on Siamese Network [4]. In our implementation, we used three convolution layers and one fully connected layer. It is observed that the deep learned features got from this model provides comparable performance with other pre-trained models considered. We also advocate the use of supervised hashing method in the existing pipeline of biometric identification system to reduce the dimensionality of

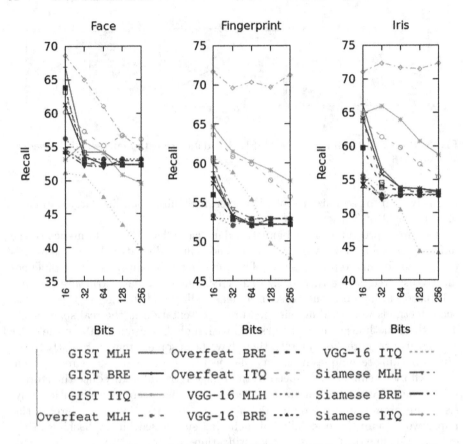

Fig. 4. Comparing hashing methods - MLH, BRE, and ITQ with respect to Bit vs Recall Performance

biometric databases. Such a suggestion is based on the thorough evaluation of various feature representations and hashing techniques for multiple biometric traits. These methods, in general use random projection to map feature vectors in real space to binary space such that similar items in real space concerning Euclidean distance mapped to objects of low Hamming distance in binary space. In the evaluation, we considered both supervised and unsupervised techniques. It is observed that supervised methods are better than unsupervised ones. Finally the performance of the methods are compared under these classes which generate a binary and non-binary representation of the data.

The paper is organized as follows. Section 2 describes the proposed approach, where we briefly covers the various hashing methods used in our experiments and also discusses the details of Siamese network based feature generation. Section 3 is devoted to experimentation and analysis. In Sect. 4 we discusses the results to form concluding remarks and finally Sect. 5 concludes our findings.

Table 1. Area under the curve values for LFW Face database for Siamese(4096D) vs Gist(512D) vs Overfeat(4096D) vs VGG-16(8192)

Face - Siamese					
Bits	16	32	64	128	256
MLH	0.29	0.50	0.63	0.75	0.82
BRE	0.23	0.37	0.51	0.6	0.69
ITQ	0.39	0.44	0.49	0.54	0.56
SKLSH	0.24	0.45	0.61	0.73	0.83
LSH	0.17	0.25	0.31	0.40	0.51
PCA	0.21	0.22	0.19	0.18	0.20
Face - Gist					
MLH	0.30	0.68	0.77	0.83	0.88
BRE	0.22	0.43	0.61	0.73	0.78
ITQ	0.34	0.40	0.49	0.51	0.53
SKLSH	0.10	0.23	0.27	0.61	0.59
LSH	0.24	0.32	0.36	0.40	0.49
PCA	0.21	0.19	0.14	0.13	0.14
Face - Overfeat					
MLH	0.35	0.68	0.80	0.83	0.85
BRE	0.26	0.42	0.62	0.74	0.79
ITQ	0.42	0.43	0.50	0.52	0.45
SKLSH	0.29	0.33	0.49	0.56	0.74
LSH	0.34	0.37	0.43	0.43	0.44
PCA	0.22	0.21	0.19	0.19	0.15
Face - VGG-16					
MLH	0.23	0.38	0.44	0.59	0.78
BRE	0.19	0.29	0.41	0.54	0.63
ITQ	0.28	0.17	0.37	0.44	0.47
SKLSH	0.15	0.16	0.19	0.32	0.39
LSH	0.15	0.16	0.26	0.33	0.40
PCA	0.24	0.21	0.20	0.20	0.18

2 Proposed Approach

The focus of this study is to evaluate whether it is possible to obtain a compact representation for multiple biometric data. We propose a modification in the existing bio-metric identification/verification pipeline. Where we feed the identification stage of such system with binary codes equivalent to the real space feature representation. For the evaluation, we conduct experiments starting with generation of both non deep and deep learned features for Face, Iris, and Fingerprint databases. The features are generated using Gist, VGG-16,

Table 2. Area under the curve values for CASIA Fingerprint database for Siamese(4096D) vs Gist(512D) vs Overfeat(4096D) vs VGG-16(8192)

Fingerprint - Siamese					
Bits	16	32	64	128	256
MLH	0.28	0.56	0.74	0.83	0.88
BRE	0.23	0.42	0.58	0.70	0.76
ITQ	0.36	0.43	0.39	0.49	0.46
SKLSH	0.29	0.31	0.55	0.72	0.80
LSH	0.29	0.39	0.43	0.47	0.49
PCA	0.22	0.19	0.19	0.19	0.20
Fingerprint - Gist					
MLH	0.28	0.56	0.71	0.81	0.87
BRE	0.26	0.45	0.57	0.68	0.75
ITQ	0.29	0.33	0.36	0.32	0.31
SKLSH	0.24	0.27	0.34	0.53	0.65
LSH	0.18	0.26	0.32	0.32	0.41
PCA	0.25	0.21	0.21	0.18	0.16
Fingerprint - Overfeat					
MLH	0.33	0.64	0.76	0.85	0.89
BRE	0.25	0.42	0.57	0.68	0.74
ITQ	0.31	0.33	0.35	0.35	0.38
SKLSH	0.25	0.28	0.31	0.51	0.62
LSH	0.20	0.29	0.32	0.34	0.40
PCA	0.23	0.25	0.22	0.20	0.17
Fingerprint - VGG-16					
MLH	0.26	0.44	0.58	0.72	0.79
BRE	0.20	0.29	0.42	0.52	0.60
ITQ	0.36	0.39	0.47	0.54	0.55
SKLSH	0.11	0.22	0.27	0.36	0.55
LSH	0.20	0.24	0.35	0.42	0.49
PCA	0.26	0.23	0.20	0.22	0.17

Overfeat (OF), and a custom designed Siamese network (SIA). The various feature dimensions obtained are 512, 8192, 4096, and 4096 respectively. Once such features are available, we explored various hashing mechanisms to get approximate nearest neighbours of the query data. We used unsupervised methods like PCA, LSH, SKLSH and ITQ. We also considered supervised methods like BRE, and MLH.

We then conduct analysis of retrieval performance for binary code lengths 16, 32, 64, 128, and 256 bits corresponding to combinations of feature

Table 3. Area under the curve values for CASIA Iris database for Siamese(4096D) vs Gist(512D) vs Overfeat(4096D) vs VGG-16(8192)

Iris - Siamese					
Bits	16	32	64	128	256
MLH	0.33	0.56	0.66	0.79	0.86
BRE	0.32	0.44	0.58	0.66	0.74
ITQ	0.40	0.38	0.42	0.44	0.45
SKLSH	0.30	0.39	0.56	0.70	0.81
LSH	0.28	0.37	0.39	0.44	0.47
PCA	0.22	0.22	0.18	0.19	0.19
Iris - Gist					
MLH	0.21	0.54	0.73	0.80	0.85
BRE	0.18	0.36	0.51	0.61	0.67
ITQ	0.59	0.60	0.65	0.65	0.67
SKLSH	0.39	0.40	0.59	0.80	0.89
LSH	0.48	0.49	0.56	0.65	0.68
PCA	0.40	0.33	0.28	0.26	0.22
Iris - Overfeat					
MLH	0.24	0.52	0.70	0.80	0.83
BRE	0.18	0.39	0.52	0.61	0.67
ITQ	0.30	0.37	0.33	0.37	0.40
SKLSH	0.18	0.25	0.39	0.54	0.65
LSH	0.20	0.28	0.38	0.33	0.37
PCA	0.25	0.21	0.22	0.18	0.15
Iris - VGG-16					
MLH	0.20	0.37	0.51	0.61	0.72
BRE	0.19	0.27	0.40	0.52	0.61
ITQ	0.31	0.32	0.33	0.40	0.38
SKLSH	0.16	0.21	0.33	0.35	0.51
LSH	0.17	0.19	0.28	0.26	0.36
PCA	0.25	0.22	0.19	0.19	0.20

representations and hashing methods. The details of the analysis is reported in the Experimentation section. In the following subsections we present a brief overview of the hashing methods and feature representations used in the experiments.

2.1 Hashing Methods

In this section, we explores the various hashing methods in the literature. An overview of hashing techniques are presented to gain sufficient insight on theses

Table 4. MAP values for comparing effect of code length on various hashing methods

Feature	MLH				BRE				ITQ			
	SIA	Gist	OF	Vgg	SIA	Gist	OF	Vgg	SIA	Gist	OF	Vgg
Face												
256	0.9	0.9	0.87	0.82	0.75	0.8	0.8	0.67	0.5	0.53	0.48	0.61
128	0.81	0.85	0.85	0.62	0.67	0.74	0.75	0.58	0.48	0.51	0.52	0.56
64	0.69	0.79	0.81	0.47	0.6	0.62	0.63	0.43	0.42	0.48	0.42	0.49
32	0.53	0.73	0.73	0.41	0.38	0.45	0.43	0.32	0.36	0.37	0.35	0.19
16	0.36	0.43	0.47	0.26	0.35	0.27	0.34	0.21	0.29	0.3	0.37	0.32
Fingerprint												
256	0.9	0.9	0.91	0.81	0.77	0.78	0.76	0.62	0.33	0.37	0.42	0.56
128	0.84	0.84	0.87	0.74	0.72	0.7	0.69	0.53	0.36	0.35	0.38	0.54
64	0.75	0.74	0.78	0.59	0.6	0.6	0.58	0.43	0.29	0.33	0.36	0.46
32	0.6	0.58	0.67	0.46	0.44	0.47	0.45	0.3	0.33	0.31	0.3	0.39
16	0.35	0.33	0.4	0.28	0.3	0.31	0.28	0.21	0.26	0.24	0.27	0.34
Iris												
256	0.89	0.89	0.86	0.75	0.75	0.7	0.7	0.64	0.31	0.54	0.43	0.54
128	0.81	0.83	0.83	0.63	0.69	0.64	0.64	0.54	0.31	0.51	0.39	0.54
64	0.7	0.77	0.73	0.53	0.57	0.54	0.54	0.42	0.31	0.48	0.33	0.44
32	0.56	0.6	0.55	0.38	0.45	0.39	0.42	0.28	0.26	0.44	0.34	0.39
16	0.36	0.31	0.32	0.22	0.34	0.25	0.22	0.2	0.3	0.46	0.26	0.32

methods that form the crux of this paper. Our main observation after empirical analysis leads to recommending the use of supervised binary hashing for the representation of biometric data.

Locality Sensitive Hashing (LSH). LSH [1,3] is an unsupervised data independent hashing method, which reduces dimensionality of input data by mapping similar items to same buckets with high probability.

Locality-Sensitive Binary Codes from Shift-Invariant Kernels (SKLSH). SKLSH [9] proposed by Raginsky and Lazebnik is also an unsupervised data independent hashing method. It uses random projection to obtain a binary encoding of data such that similar data points map to binary strings with low Hamming distance.

Iterative Quantization (ITQ). Iterative Quantization is a simple and efficient dimensionality reduction scheme, proposed by Gong and Lazebnik [2] to reduce the quantization error by mapping the high dimensional data to the vertices of a binary hypercube (zero-centered).

Binary Reconstructive Embedding (BRE). Binary Reconstructive Embedding (BRE) is a supervised hashing method proposed by Kulis and Darell [5]. The method uses the learning of hash functions that minimize reconstruction error between the original distances and the Hamming distances of the corresponding binary embeddings. A scalable coordinate-descent algorithm is used for the proposed hashing objective to learn hash functions in a variety of settings efficiently.

Minimum Loss Hashing for Compact Binary Codes (MLH). Norouzi and Blei proposed Minimum loss hashing [7] which is a supervised binary hashing technique that uses random projections to map high-dimensional input into binary codes. It assigns a 1, if the bit corresponding to the input is on one side of the hyperplane and 0, if it is on the other side. Then a hinge-like loss function in SVM, which based on some threshold ρ bits in the Hamming space assign a cost to a pair of binary codes and a similarity label. Finally, it learns a parameter matrix w which maps high dimensional inputs to binary codes by minimizing the empirical loss over training points.

2.2 Feature Representations

We implemented a custom Siamese neural network (SIA) based on [4]. The one leg of the network contains three convolution layers and two fully connected layers. The kernel size is 3×3 with stride of 1 and no padding. The ReLU activation function is applied after each convolution layer. The first two convolution layer is also followed by max pooling layer. In the fully connected layer, we flatten the output of convolution layer and then sigmoid function is used to obtain a 4096 dimension feature vector. Another sigmoid function is also applied before we take a single-valued vector output. We feed two images of dimension 105×105 in parallel into two legs of the Siamese network to detect the similarity between them. We randomly feed a pair of similar or a pair of dissimilar images during every iteration. The output from each leg of the network is compared to see if the images are similar or not. We used **Binary cross entropy with logits** as the loss function. We used a learning rate of 0.0001 and Adam as the optimizer with a weight decay of $1-e5$. This experiment could achieve an accuracy above 96% for Face and Iris case, but accuracy in the case of the Fingerprint was about 90 (Fig. 5).

We also extracted feature descriptors using Overfeat (OF) [11] CNN, and torch CNN using VGG-16 [12] training model. The implementation provided by CILVR lab at New York University [11] are used to extract Overfeat features. We used the torch CNN with VGG-16 training model which runs only on CPU. We took the output from the fc7 layer of both CNNs, which gives 4096 and 8192 dimension feature vectors respectively.

We used the original implementation provided by Olivia and Torrabla [8] for extracting Gist features of 512 dimensions.

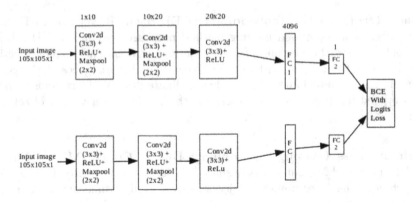

Fig. 5. Proposed Siamese network

3   Experimentation

We performed our experiments on popular databases of the Face, Iris, and Fingerprint which ensure variation in the database and across the databases relevant to the feature points. Some of the images in the Face database are profile picture in the case of face images. Some of the iris images taken with spectacles on and some fingerprint images are rotated. Each database has undergone three traversals of training and query traversal. The Face database we use is Labeled Faces in the Wild (LFW) which consists of 13,233 images. The Iris database we use is CASIA-Iris-Thousand version 4.0 of 1000 subjects. The CASIA Fingerprint image database Version 5.0 of 500 subjects is used as the Fingerprint database. Both of these databases consist of 20,000 images.

We divide each data set into 1000 training samples and 3000 testing samples. On each training set, the Euclidean distance for each data point is computed to find their 100 ground-truth neighbours. Then we compute precision and recall statistics during testing using the ground-truth neighbours and non-neighbours.

3.1   Analysis of Feature Vector Representations

We analyze the retrieval performance concerning feature representation of different modalities, using Precision-Recall for the MLH and BRE which are the supervised hashing techniques. We also compare the Precision-Recall performance of unsupervised method ITQ to establish the superior performance of supervised methods over the unsupervised methods. We evaluate the feature representations obtained from Siamese network, Gist, Overfeat, and VGG-16 for bit sizes ranging from 16 to 256 bits. The Precision-Recall graphs for this comparison are provided in Figs. 1, 2 and 3. We have omitted the curves of LSH, SKLSH, and PCA to avoid cluttering. Also, we have not included the Precision-Recall curves for 16 bit and 32 bit representation to save the space. It is observed that the performance is not good for these code lengths in general. We have also provided mean average precision (MAP) values in Table 4. We observe that the

Siamese feature representation is performing better for Face and Iris concerning MAP values for a bit size of 256. Overfeat performed slightly better in the case of Fingerprint for a bit size of 256. The Gist representation is also giving comparable performance. The VGG-16 representation provides a slightly inferior performance, which is more visible (from Tables 3 and 4) for the lower bit sizes up to 64 bits of all the modalities. We observed that consistent retrieval performance achieved across all the modalities, with Siamese, Gist or Overfeat representation.

3.2 Analysis of Hashing

We can infer from Fig. 4, and Tables 1, 2 and 3 most of the unsupervised methods are inferior to supervised methods, such as MLH and BRE. We also compare the unsupervised method ITQ as it was performing better in 16 bit and 32 bit case of Iris. If we ignore the lower bit cases (16 and 32) of Iris, in all other cases MLH was performing better. Among the unsupervised methods, SKLSH was performing better for the 128 bit and 256 bit case and ITQ otherwise. It is recommended to use the supervised method MLH with 128 bit or 256 bit to achieve a better retrieval performance across all modalities. Some of the unsupervised methods show better performance in the Bit versus recall curves. But their precision performance is poor compared to supervised methods as obvious from the Tables 1, 2, 3 and 4. This means that the relevant items retrieved may be containing more false items.

3.3 Biometric Wise Analysis of Methods

All biometric databases retrieved with better accuracy by supervised methods, especially the MLH, compared to unsupervised methods except for 16 and 32 bit cases of Iris database. It is evident from Tables 1, 2, 3 and 4, the Fingerprint gave the maximum accuracy with 256 bit size with overfeat feature representation.

3.4 Computation Factors

The training time for supervised techniques was taking around 12 to 14 h on core i7 desktop with 16 GB RAM. The training time was almost uniform regardless of the modalities or feature representation in the supervised setting. But on an i7 machine with 8 GB RAM it took nearly two days to finish the training. The unsupervised techniques were taking a maximum of a couple of minutes for the entire process for all modalities. The feature vector generation with Gist took the least time, an hour, with VGG-16 it took nearly 6 h, and with Overfeat takes more than 24 h. Siamese network took around 2 h for Iris and Face and 6 h for the Fingerprint case when running on GTX 1080 Ti.

4 Discussion

Experimentation results show that MLH supervised method works consistently well with all biometric databases for binary code size from 64 bits to 256 bits. So for any of the Face, Fingerprint and Iris databases, MLH can be the best choice. In the case where (16 and 32 bit cases of Iris) unsupervised methods are superior, but the overall accuracy is less in those cases. So it is recommended to use MLH on the combination of the above three biometric databases sets, with a binary code size of 256 bits.

If we ignore computational limitations regarding feature vector generation time or training time, then MLH provides the best accuracy over Siamese and Overfeat feature representations for a bit size of 256 bits. Both methods were performing comparably well. We also observed that maximum accuracy obtained for Face and Iris databases with Siamese and Gist. In the case of the Fingerprint database, Overfeat provided maximum accuracy.

As we have seen previously, supervised methods are performing consistently well across all data sets for 128 and 256 bit sizes regardless of the computational need and biometric modality. Then if we have to find a trade-off between accuracy and storage, then 128 bit could be the best choice. This is recommended because, with 128 bit, accuracy is closer to that of 256 bits, while it needs less storage size compared to 256 bit case.

From Tables 1, 2, 3 and 4 and Figs. 1, 2 and 3, we found that MLH over Siamese features for Face and Iris and MLH over Overfeat feature representation of the Fingerprint database performed most accurate for a bit size of 256.

So if we consider the computational time for preparing the feature representation, then it would be better to choose Siamese feature representation. We suggest this because it gives better or comparable performance to the pre-trained feature representation for all data sets, for bit sizes of 256 bits.

5 Conclusions

Our experimentation and analysis show that MLH supervised hashing method performs consistently better than unsupervised methods for all bit lengths except 16 bit case of Iris database across all feature representations. So it would be ideal to use a setting where supervised hashing employed for multimodal biometric data retrieval with feature representations being either Siamese or Overfeat and a bit size of 256. Gist features are also found to be competitive with deep learning based general features.

References

1. Charikar, M.S.: Similarity estimation techniques from rounding algorithms. In: Proceedings of the Thirty-Fourth Annual ACM Symposium on Theory of Computing, pp. 380–388. ACM (2002)

2. Gong, Y., Lazebnik, S.: Iterative quantization: a procrustean approach to learning binary codes. In: 2011 IEEE Conference on Computer Vision and Pattern Recognition (CVPR), pp. 817–824. IEEE (2011)
3. Indyk, P., Motwani, R.: Approximate nearest neighbors: towards removing the curse of dimensionality. In: Proceedings of the Thirtieth Annual ACM Symposium on Theory of Computing, pp. 604–613. ACM (1998)
4. Koch, G., Zemel, R., Salakhutdinov, R.: Siamese neural networks for one-shot image recognition. In: ICML Deep Learning Workshop, vol. 2 (2015)
5. Kulis, B., Darrell, T.: Learning to hash with binary reconstructive embeddings. In: Advances in Neural Information Processing Systems, pp. 1042–1050 (2009)
6. Ngo, D.C., Teoh, A.B., Goh, A.: Biometric hash: high-confidence face recognition. IEEE Trans. Circuits Syst. Video Technol. 16(6), 771–775 (2006)
7. Norouzi, M., Blei, D.M.: Minimal loss hashing for compact binary codes. In: Proceedings of the 28th International Conference on Machine Learning (ICML 2011), pp. 353–360. Citeseer (2011)
8. Oliva, A., Torralba, A.: Modeling the shape of the scene: a holistic representation of the spatial envelope. Int. J. Comput. Vis. 42(3), 145–175 (2001)
9. Raginsky, M., Lazebnik, S.: Locality-sensitive binary codes from shift-invariant kernels. In: Advances in Neural Information Processing Systems, pp. 1509–1517 (2009)
10. Rathgeb, C., Uhl, A.: Iris-biometric hash generation for biometric database indexing. In: 2010 20th International Conference on Pattern Recognition (ICPR), pp. 2848–2851. IEEE (2010)
11. Sermanet, P., Eigen, D., Zhang, X., Mathieu, M., Fergus, R., LeCun, Y.: OverFeat: integrated recognition, localization and detection using convolutional networks. arXiv preprint arXiv:1312.6229 (2013)
12. Simonyan, K., Zisserman, A.: Very deep convolutional networks for large-scale image recognition. arXiv preprint arXiv:1409.1556 (2014)
13. Sutcu, Y., Sencar, H.T., Memon, N.: A secure biometric authentication scheme based on robust hashing. In: Proceedings of the 7th Workshop on Multimedia and Security, pp. 111–116. ACM (2005)
14. Tulyakov, S., Farooq, F., Govindaraju, V.: Symmetric hash functions for fingerprint minutiae. In: Singh, S., Singh, M., Apte, C., Perner, P. (eds.) ICAPR 2005. LNCS, vol. 3687, pp. 30–38. Springer, Heidelberg (2005). https://doi.org/10.1007/11552499_4

Pose Estimation for Distracted Driver Detection Using Deep Convolutional Neural Networks

Siddhesh Thakur, Bhakti Baheti$^{(\boxtimes)}$, Suhas Gajre, and Sanjay Talbar

Shri Guru Gobind Singhji Institute of Engineering and Technology,
Nanded, Maharashtra, India
{2015BCS066,bahetibhakti,ssgajre,sntalbar}@sggs.ac.in

Abstract. Distracted driver has been a major issue in today's world with more than 1.25 million road incidents of fatality. Almost 20% of all the vehicle crashes occur due to distracted driver. We attempt to create a warning system which will make the driver attentive again. This paper focuses on a simple yet effective Convolutional Neural Network technique which can help us to detect if the driver is safely driving or is distracted which is a binary classification task. It would help in improving the safety measures of the driver and vehicle. We propose two techniques for distracted driver detection achieving state of the art results. We achieve an accuracy of 96.16% for the 10 class classification. We propose to deconstruct the problem into a binary classification problem and achieve an accuracy of 99.12% for the same. We take advantage of recent techniques of transfer learning combined with regularization techniques to achieve these results.

Keywords: Distracted driver detection · Driver pose estimation · ADAS · Deep learning

1 Introduction

One of the major issues in self-driving technology as well non automated driving has been driver distraction. According to a report of World Health Organization (WHO) [18], more than 1.25 million fatalities and 20–50 million incidents of injuries occurred due to this issue worldwide in year 2017 and its expected to be 5th leading cause of death worldwide by the year 2030 across 80 countries. Reports by National Highway Traffic Safety Administrator (NHTSA) [11] state that this number of fatalities has been increasing worldwide with a major increase in poorer countries and the major cause of this has been the usage of cell phones during driving. According to the National Crime Research Bureau (NCRB), India has one of the highest rates of on-road fatalities in the world. 0.15 million deaths took place in 2016 alone out of which 0.135 million were due to negligent and distracted driving, which keeps on increasing as to date.

NHTSA describes distracted driving as "any activity that diverts attention of the driver from the task of driving" which is majorly classified into cognitive,

© Springer Nature Singapore Pte Ltd. 2019
C. Arora and K. Mitra (Eds.): WCVA 2018, CCIS 1019, pp. 102–114, 2019.
https://doi.org/10.1007/978-981-15-1387-9_9

visual and manual. It could range from activities such as Talking to passenger, Looking behind, Moving Object, Changing Radio Stations, Eating or Drinking, Using cellphones etc. In our research work, we focus on detecting whether the driver is distracted and warning him against it and not on how the driver is distracted. This could help us in creating facilitatory functions which would alert the systems about the driver distraction and take preventive measures to warn the driver about it. And if no action was taken by the driver, the system would slow down or stop the vehicle to avoid accidents and help in improving the safety of drivers and passengers.

In this paper, we propose an approach to answer the question of driver distraction by implementing a Convolutional Neural Network which provides a good detection rate for detection of distracted driver and help in improving the driver and road safety. Remainder of this paper is structured as follows: Sect. 2 briefs about recent research in driver state monitoring. Dataset is described in Sect. 3. Section 4 contains details of technical approach. Results are discussed in Sect. 5. Section 6 presents conclusion and future scope.

2   Related Work

We get a small review on recent and relevant work from literature for the pose estimation of distracted driver. According to a report [11], use of cellphones is one of the biggest cause for driver distraction. So the research was conducted for cell phone detection while driving by Zheng et al. [20]. This work was improved in 2015 by Das et al. [3] by creating a dataset for hand detection in the car and achieved an average precision of 70.09% using ACF object detector. Similar work was also depicted by Seshadri et al. [13] for cell phone usage detection and classification using Supervised Descent Method and HoG with AdaBoost Classifier for achieving 93.9% accuracy at 7.5fps. A higher accuracy of 94.2% was achieved beating the state of the art models by the method of Faster RCNN by Le et al. [8] on the above dataset. It majorly focussed on detection of the face and hand regions for detecting cell-phone usage.

Zhao et al. [21] created a more inclusive dataset with obtaining a view of the driver from side and considering various activities such as: Eating, Talking on a cell phone, operating shift gear handle and driving safe. They were able to achieve 90.5% classification accuracy using a contourlet transform and random forest and they boosted their system using PHOG and MLP that yielded an accuracy of 94.75% [22]. So continuing the challenge, Yan et al. [19] proposed a method implementing CNN and achieved a 99.78% accuracy. Ohn-bar et al. [12] also proposed an ensembled fusion of classifiers to achieve better results.

These datasets focused upon the limited distractions that could occur. Also most of them are inaccessible to most researchers as datasets are not publicly available. In 2016, State Farm [4] hosted a competition for distracted driver detection on 10 different postures and this was the first database which was made publicly available. Many approaches were proposed based on traditional methods such as SIFT, HoG with SVM and BoW. However, CNN's were the most effective technique in achieving state of the art results [6].

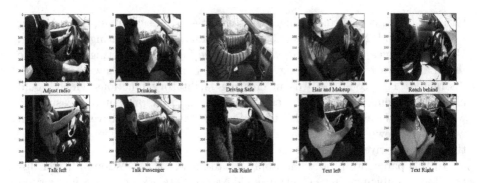

Fig. 1. Ten classes of distracted driver detection dataset

In 2017, a new dataset [1] was created which was similar to State Farm's dataset for the distracted driver and was publicly available. Authors proposed to use an Ensemble of 5 different CNN to achieve high accuracy which made it computationally expensive and too complex to be used in real time which is very important for autonomous driving. Baheti et al. [2] also proposed a CNN based approach for distraction detection and classification on this dataset. It is to be observed that all the approaches from literature are based on detecting and classifying the distraction. However, detecting whether the driver is distracted or not and alerting him is more important than identifying the cause of distraction. Hence in our research, we propose to work on driver distraction as binary classification problem instead of ten class classification and achieve good accuracy.

3   Dataset Description

The AUC distracted driver dataset [1] used in this paper consists of primarily 10 classes as ($c0$) Adjust Radio, ($c1$) Drinking, ($c2$) Driving Safe, ($c3$) Hair and Makeup, ($c4$) Reach Behind, ($c5$) Talking using Left hand on the cellphone, ($c6$) Talking to Passenger, ($c7$) Talk using right hand on the cellphone, ($c8$) Texting with left hand, and ($c9$) Texting with right hand. Figure 1 shows sample images of each class resized to 299 × 299. Videos were shot in 4 different cars and 31 participants were called up from 7 different countries. The dataset consists of total of 17308 images of 1920 × 1080 resolution divided into the training set (12977) and test set (4331). We use this dataset as it is for ten class classification problem. Figure 2(a) shows original distribution of the dataset amongst 10 classes. As our major issue of Driver distraction should be addressed first which makes the problem statement a binary classification instead of a categorical classification. So, we propose to only detect driver's distraction and divide the dataset into only two classes viz. (i) Driving Safely and (ii) Distracted Driver Detected as a binary classification problem. Data distribution of the same is shown in Fig. 2(b). Since This might seem to further simplify the quintessential task of our problem statement, since binary classification may come to be seen as easier than the

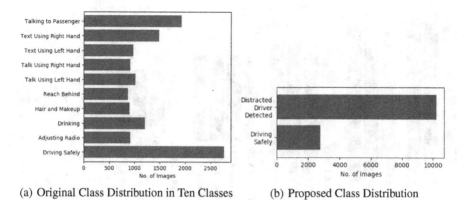

(a) Original Class Distribution in Ten Classes   (b) Proposed Class Distribution

Fig. 2. Category-wise dataset distribution

categorical classifications for the modern CNN architectures which are available. It can be observed from Fig. 2(b) that we face a massive class imbalance for our problem statement which makes the task more difficult.

4 Our Approach

Convolutional Neural Network (CNN) is a class of Deep Neural Networks mostly used in analysing images. CNNs have wide range of applications like image classification, object detection, action recognition etc. Motivated by the performance of CNN in various tasks of computer vision, we explore VGG and InceptionResNetV2 architecture for detecting driver distraction.

4.1 Network Architectures

VGG16 Architecture: VGG16 is a CNN architecture proposed by Simonyan and Zisserman [14] which achieved good performance in ImageNet classification as well as localisation challenge. We trained the classical VGG16 architecture initially to address the problem. The VGG16 architecture as shown in Fig. 3 consists of 13 sequential layers of connecting Convolutional Layers of 3×3 filter size, 2×2 max-pooling operation with a stride of 2 with basic activation function ReLU. Then these layers are followed by a simple flatten layer followed by two dense layers of 4096 neurons which is further followed by the Softmax layer with dimension of $number\,of\,classes$ N. In our case $N = 10$ for categorical classification and $N = 2$ for binary classification.

InceptionResnetV2: We also explored InceptionResnetV2 [16] architecture which is another recent and powerful technique of classification. It is basically a combination of InceptionV3 and ResNet with 50 layers. Figures 4 and 5 illustrates the concept of inception module and residual block respectively. It consists of

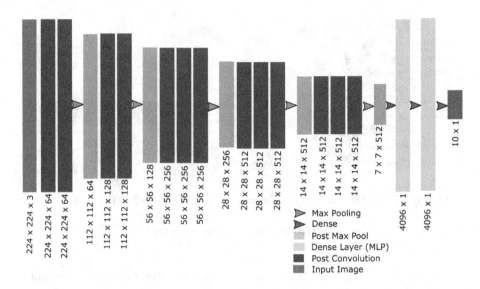

Fig. 3. VGG16 architecture for 10 class classification

Inception Modules in the first half of layers followed by a residual block. We train the network for distracted driver detection by applying concept of transfer learning.

Inception Module: Inception Module is combination of a layer of distributed convolution operations and max-pooling 3×3 acts as a layer to zoom inside the filters and understand the deeper spectrum of the layer [17]. The 1×1 convolution is a key as it helps in dimensionality reduction of its feature map. This is massively helping in reducing the computational capacity of the network as described in Fig. 4.

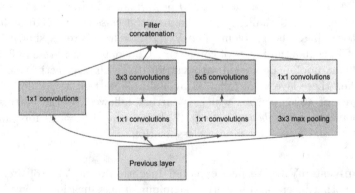

Fig. 4. Understanding Inception module in a single layer [17]

ResNet Module: The skip connection was introduced in the ResNet [5] as it helps in recollecting the knowledge of older layers. Considering x as the input and our convolution operations can be considered as $f(x)$, then we concatenate the original input with this convolution operation as $f(x) + x$ which helps in not forgetting the vital information from above layers.

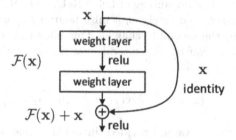

Fig. 5. Understanding ResNet module in a single layer [5]

4.2   Regularization Techniques and Fine Tuning

To improve the performance of networks and reduce overfitting, we imply various regularization techniques. We fine tune the VGG16 architecture by replacing the activation function with LeakyReLU, also introducing Batch Normalization after every layer to boost convergence, add dropout and weight regularization to avoid overfitting to the training data which we will see in the following sections.

Dropout: We apply the technique of dropout as proposed by Hinton et al. [15] which can be understood as following. Consider a simple Neural Network with L number of hidden layers, Let $l \in \{1,...,L\}$ be the index of the hidden layers of the Neural network. Let $z^{(l)}$ denote the vector of input into layer l, $y^{(l)}$ denote the vector of outputs from layer l ($y^{(0)} = x$ is the input). $W^{(l)}$ and $b^{(l)}$ are the given weights and bias for the appropriate layer l. The feed forward operation of a standard neural network can be simply described as (for $l \in \{1,...,L-1\}$ and any hidden unit i) as shown in Eq. 1.

$$z_i^{(l+1)} = w_i^{(l+1)} y^l + b_i^{l+1}$$
$$y_i = f(z_i^{(l+1)}), \tag{1}$$

where f is any particular activation function. For example $f(x) = tanh(x)$ with dropout, the feed-forward operation becomes,

$$r_j^l \sim Bernoulli(p)$$
$$\tilde{y}^{(l+1)} = r^l \star y^{(l)}$$
$$z_i^{l+1} = w_i^{l+1} \tilde{y}^l + b_i^{(l+1)} \tag{2}$$
$$y_i^{(l+1)} = f(z_i^{(l+1)})$$

It helps us in reducing the overfitting by regularizing the network by randomly dropping out/turning off some neurons. It reduces interdependent learning of neurons which optimizes the neuron for the task.

LeakyReLU: Among the various activation functions, we choose to implement LeakyReLU [9]. The major advantage of LeakyReLU over ReLU is during Backpropagation. It does not completely shut off a neuron and updates its gradient weights negatively to the incoming change. This allows having better gradient updates than ReLU. It can be understood by a simple mathematical equation shown in Eq. 3.

$$f(x) = \left\{ x \ \epsilon \ x > 0; \alpha x \ \epsilon \ x < 0, 0 < \alpha < 1 \right\} \tag{3}$$

where α is the parametric value provided during the time of compilation, and by default we choose it to be $\alpha = 0.1$ by default. It is fast and efficient in nature in terms of convergence which has made it popular in recent times.

Batch Normalization: Batch Normalization [7] is a simpler technique of boosting the convergence rate of the model by performing the task of normalization over every mini batch after every convolutional layer. This helps in boosting the gradient descent since data normalization helps in finding the descent slope in an easier way. Since Normalization comes with mean shifting, we can be sure that the data isn't disoriented and becomes correlated with learning better.

L_2 Regularization: L_2 weight regularization [10] is also known as Ridge Regression. It adds a squared magnitude of coefficient as a penalty term to the kernel losses as shown in Eq. 4. Regularization is a very important technique in machine learning to prevent overfitting. Mathematically speaking, it adds a regularization term in order to prevent the coefficients to fit too perfectly over the training data.

$$w^* = argmin_w - \sum_j \left(t(x_j - \sum_i w_i h_i(x_j)) \right)^2 + \lambda \sum_{i=1}^{k} w_i^2 \tag{4}$$

where there is a squared penalty term to help with the kernel regularization.

Transfer Learning with Pre-trained Weights: Transfer learning is a technique which helps in transferring the knowledge of one dataset to another. In transfer learning, a neural network is previously trained on some other dataset and then the weights are finetuned with respect to our application. This technique helps in the faster convergence of networks as well as producing better results since models are not trained from scratch. It also helps in delivering a better accuracy to the network.

Optimizer SGD: The Stochastic Gradient Descent optimizer is a great choice during transfer learning since unlike other learning algorithms, it is much more stable and the loss can be controlled very easily. It can also be boosted with some Momentum and Nesterov property. Equation 5 shows the process of updating weights in backpropagation with SGD.

$$w := w - \eta \nabla Q_i(w) + \alpha \Delta w \tag{5}$$

where the parameter w which minimizes $Q(w)$ is to be estimated, and η is a step size which is called as learning rate and α is the momentum that is passed as a hyperparameter.

Cross-Entropy Loss Function: For finding the loss in the classification, we choose the cross entropy loss function. It creates a function that evaluates the compatibility in a prediction and its related ground truth label by minimizing the error shown in Eq. 6.

$$E = \frac{1}{m} \sum_{i=1}^{m} \sum_{k=1}^{c} y_i^{*(k)} log \left(y_i^k \right) \tag{6}$$

where, $y_i^{*(k)}$ and $y_i^{(k)}$ are respectively the ground-truth label and the predicted output of the i_{th} image of the k_{th} class with m training images. The total number of classes is given as c; in this case, $c = 10$ for 10 class experiment and $c = 2$ for binary class experiment. The loss is then back-propagated to update the network parameters with SGD the optimizer.

5   Experimentation and Results

We explore two networks viz. VGG16 and InceptionResnetV2 for the task of distracted driver detection. The networks were trained on an Intel Core $i7$ Processor with 16 GB RAM and a $P5000$ GPU with 2560 Cuda cores and 16 GB of VRAM and with a TensorFlow back-end accompanied by Keras front end for faster development of the 10 class classification problem. For the binary class classification, we used MATLABR12018a with CUDA backend and then the Binary cross-entropy was used as the loss function with metrics set to accuracy. We used the Stochastic Gradient Descent optimiser with learning rate of 0.0001 and with a momentum of 0.9 with *Nesterov* per epoch decay rate of 10^{-6} for both of them. Batch size was set to 64 and training was carried out for 300 epochs for the 10 class classification. The batch size and epochs were set to 32 and 25 for the binary classification problem.

We observe that both the models tend to overfit on the training set producing a 100% accuracy on the training set and in order to reduce the over-fitting, we chose to add regularization methods which significantly improved our performance. We experimented with several hyperparameters such as varying dropout,

Table 1. Confusion matrix for 10 class classification problem

	c0	c1	c2	c3	c4	c5	c6	c7	c8	c9
c0	**317**	0	4	0	3	0	2	0	0	0
c1	3	**322**	14	0	0	0	2	0	0	0
c2	2	2	**879**	2	0	5	4	7	3	15
c3	9	2	0	**474**	5	2	0	0	1	1
c4	0	0	0	10	**295**	0	1	0	0	0
c5	6	0	0	0	0	**299**	0	0	0	0
c6	3	0	0	1	0	2	**394**	0	2	1
c7	9	0	1	0	0	0	0	**288**	1	2
c8	11	0	0	0	0	0	4	2	**271**	2
c9	9	0	0	0	0	0	7	0	1	**626**

regularization strengths, different weight initialization, activation functions, optimizers and after rigorous experimentation, we chose a set out of them based on the mixture which gave a maximum accuracy. The set was chosen as L_2 regularization strength of 0.03 and a Linearly increasing dropout from 0.3 to 0.5 after every sub convolutional block. Weights are initialised with orthogonal weight initialization and Leaky ReLU activation function is used with $\alpha = 0.3$. Weights are updated with stochastic gradient descent optimiser.

Performance of distracted driver detection is evaluated on 10 class classification problem as well as binary classification problem. We freeze the VGG16's first 3 layers and update the weights of remaining layers in backpropagation. Results of VGG16 architecture for 10 class classification are evaluated and confusion matrix of the same is depicted in Table 1. We used pre-trained ImageNet weights for weight initialization for deploying the models before training. With the regularized version of VGG16 network, 10 class classification accuracy of 96.16% is achieved and processes 42 fps on an average for categorical classification.

Next, we evaluate the performance of distraction detection with only two classes as described earlier. Since we assume that driver can be distracted in multiple scenarios but the distraction could yield to an accident no matter however insignificant it may be, so we propose to use second network with a much higher accuracy for the pose estimation and also having a computationally less expensive model. After experimentation on binary classification, we found out that VGG16 tends to poorly handle class imbalance which causes the network to be biased. Network automatically gets biased to the class with higher number of samples and doesn't learn well about the class having less samples. Retraining VGG16 with class balancing techniques like weighted backprop could be replaced with application of fine tuning. So instead of trying to deal with class imbalance techniques, we preferred to choose InceptionResNetV2 and fine-tune its top layer instead which helped in improving the performance.

Table 2. Summary of comparison with other earlier approaches from literature for 10 classes

Model	Source	No. of parameters	Accuracy
AlexNet [1]	Original	62M	93.65%
	Skin segmented	62M	93.60%
	Face	62M	84.25%
	Hands	62M	89.52%
	Face+Hands	62M	86.68%
InceptionV3 [1]	Original	62M	95.17%
	Skin segmented	62M	94.57%
	Face	62M	88.82%
	Hands	62M	91.62%
	Face+Hands	62M	90.88%
Real-time system [1]		–	94.29%
Majority voting ensemble [1]		–	95.77%
GA-weighted ensemble [1]		–	95.98%
VGG16 [2]		138M	96.31%
Regularized VGG16		138M	**96.16%**
InceptionResnetV2		54M	94.93%

Table 3. Confusion matrix for the binary classification

Position	Safe driver	Distracted driver
Safe driver	**881**	13
Distracted driver	25	3412

We observe that InceptionResnet V2 performs better on binary classification whereas VGG16 with ImageNet pre-trained weights performs better for the 10 class classification. We fine tuned the InceptionResnetV2 for binary classification with L_2 regularization by adding a single node after the final layer and keep the whole network frozen. Fine tuning the InceptionResNetV2 produces categorical classification accuracy of 94.93%. Results are compared with state-of-art networks from literature in Table 2. It processes 61 fps for binary classification. Table 3 shows the confusion matrix for binary classification and we can see that higher accuracy for detecting whether the driver is safely driving or not is achieved. We evaluate the performance of binary classification i.e safe and unsafe driving in terms of Accuracy, Sensitivity, Specificity and Recall which are defined as follows.

Table 4. Binary classification accuracy with proposed method

Model	Accuracy	Sensitivity	Specificity	Precision
InceptionResnetV2	**99.12%**	**0.9854**	**0.9927**	**0.9724**
Regularized VGG16	99.07%	0.9843	0.9924	0.9712

$$Accuracy = \frac{TP + TN}{TP + FP + TN + FN} \tag{7}$$

$$Sensitivity = \frac{TP}{TP+FN} \tag{8}$$

$$Specificity = \frac{TN}{TN + FP} \tag{9}$$

$$Precision = \frac{TP}{TP + FP} \tag{10}$$

Here, TP, FP, TN and FN stand for True Positives, False Positives, True Negatives and False Negatives respectively. Table 4 shows that both the networks achieve good classification accuracy.

6   Conclusion and Future Work

Distracted driver is one of the leading issues in increased number of road crashes and accidents. Hence detection of the driver pose can help us in deciding whether he is safely driving or not in order to save lives. We develop a CNN based system for pose estimation of the driver and alerting him if he is distracted. We focus on detecting the driver distraction and not focusing on by which activity he is distracted. We imply various regularization techniques to reduce overfitting. We achieved 96.16% accuracy for the 10 class classification and 99.12% for binary classification. Results are compared with state-of-art methods from literature for driver pose estimation. This system processes 42 fps for 10 class problem and processes 63 fps for the binary classification on an NVIDIA P5000 GPU with 16 GB VRAM. As an extension of this work, we are planning to create a computationally smaller and efficient network which is faster in nature and is more deployable practically.

References

1. Abouelnaga, Y., Eraqi, H.M., Moustafa, M.N.: Real-time distracted driver posture classification. CoRR abs/1706.09498 (2017). http://arxiv.org/abs/1706.09498
2. Baheti, B., Gajre, S., Talbar, S.: Detection of distracted driver using convolutional neural network. In: The IEEE Conference on Computer Vision and Pattern Recognition (CVPR) Workshops, June 2018

3. Das, N., Ohn-Bar, E., Trivedi, M.M.: On performance evaluation of driver hand detection algorithms: challenges, dataset, and metrics. In: 2015 IEEE 18th International Conference on Intelligent Transportation Systems, pp. 2953–2958, September 2015. https://doi.org/10.1109/ITSC.2015.473
4. Farm, S.: State farm distracted driver detection (2016). https://www.kaggle.com/c/state-farm-distracted-driver-detection
5. He, K., Zhang, X., Ren, S., Sun, J.: Deep residual learning for image recognition. CoRR abs/1512.03385 (2015). http://arxiv.org/abs/1512.03385
6. Hssayeni, M., Saxena, S., Ptucha, R., Savakis, A.: Distracted driver detection: deep learning vs handcrafted features 2017, 20–26 (2017)
7. Ioffe, S., Szegedy, C.: Batch normalization: Accelerating deep network training by reducing internal covariate shift. CoRR abs/1502.03167 (2015). http://arxiv.org/abs/1502.03167
8. Le, T.H.N., Zheng, Y., Zhu, C., Luu, K., Savvides, M.: Multiple scale faster-RCNN approach to driver's cell-phone usage and hands on steering wheel detection. In: 2016 IEEE Conference on Computer Vision and Pattern Recognition Workshops (CVPRW), pp. 46–53, June 2016. https://doi.org/10.1109/CVPRW.2016.13
9. Maas, A.L.: Rectifier nonlinearities improve neural network acoustic models (2013)
10. Ng, A.Y.: Feature selection, l1 vs. l2 regularization, and rotational invariance. In: Proceedings of the Twenty-first International Conference on Machine Learning, ICML 2004, p. 78. ACM, New York (2004). https://doi.org/10.1145/1015330.1015435
11. NHTSA: National highway traffic safety administration traffic safety facts. https://www.nhtsa.gov/risky-driving/distracted-driving/
12. Ohn-Bar, E., Martin, S., Tawari, A., Trivedi, M.M.: Head, eye, and hand patterns for driver activity recognition. In: 2014 22nd International Conference on Pattern Recognition, pp. 660–665, August 2014. https://doi.org/10.1109/ICPR.2014.124
13. Seshadri, K., Juefei-Xu, F., Pal, D.K., Savvides, M., Thor, C.P.: Driver cell phone usage detection on strategic highway research program (SHRP2) face view videos. In: 2015 IEEE Conference on Computer Vision and Pattern Recognition Workshops (CVPRW), pp. 35–43, June 2015. https://doi.org/10.1109/CVPRW.2015.7301397
14. Simonyan, K., Zisserman, A.: Very deep convolutional networks for large-scale image recognition. CoRR abs/1409.1556 (2014). http://arxiv.org/abs/1409.1556
15. Srivastava, N., Hinton, G., Krizhevsky, A., Sutskever, I., Salakhutdinov, R.: Dropout: a simple way to prevent neural networks from overfitting. J. Mach. Learn. Res. 15, 1929–1958 (2014). http://jmlr.org/papers/v15/srivastava14a.html
16. Szegedy, C., Ioffe, S., Vanhoucke, V.: Inception-v4, inception-resnet and the impact of residual connections on learning. CoRR abs/1602.07261 (2016). http://arxiv.org/abs/1602.07261
17. Szegedy, C., et al.: Going deeper with convolutions. CoRR abs/1409.4842 (2014). http://arxiv.org/abs/1409.4842
18. WHO: World health organization global status report on road safety 2015 (2015). https://www.who.int/violence-injury-prevention/road-safety-status/2015/en/. Accessed 03 Apr 2018
19. Yan, C., Coenen, F., Zhang, B.: Driving posture recognition by convolutional neural networks. IET Comput. Vis. 10(2), 103–114 (2016). https://doi.org/10.1049/iet-cvi.2015.0175
20. Zhang, X., Zheng, N., Wang, F., He, Y.: Visual recognition of driver hand-held cell phone use based on hidden CRF. In: Proceedings of 2011 IEEE International Conference on Vehicular Electronics and Safety, pp. 248–251, July 2011. https://doi.org/10.1109/ICVES.2011.5983823

21. Zhao, C.H., Zhang, B.L., He, J., Lian, J.: Recognition of driving postures by contourlet transform and random forests. IET Intell. Transp. Syst. **6**(2), 161–168 (2012). https://doi.org/10.1049/iet-its.2011.0116

22. Zhao, C.H., Zhang, B.L., Zhang, X.Z., Zhao, S.Q., Li, H.X.: Recognition of driving postures by combined features and random subspace ensemble of multilayer perceptron classifiers. Neural Comput. Appl. **22**(1), 175–184 (2013). https://doi.org/10.1007/s00521-012-1057-4

AECNN: Autoencoder with Convolutional Neural Network for Hyperspectral Image Classification

Heena Patel and Kishor P. Upla[✉]

Sardar Vallabhbhai National Institute of Technology, Surat, Gujarat, India
hpatel1323@gmail.com, kishorupla@gmail.com

Abstract. This paper addresses an approach for classification of hyperspectral imagery (HSI). In remote sensing, the HSI sensor acquires hundreds of images with very narrow but continuous spectral width in visible and near-infrared regions of the electromagnetic (EM) spectrum. Due to the nature of data acquisition with contiguous bands, the HS images are very useful in classification and/or the identification of materials present in the captured geographical area. However, the low spatial resolution and large volume of HS images make the classification of those images as a challenging task. In the proposed approach, we use an autoencoder with convolutional neural network (AECNN) for classification of HS image. Pre-processing procedure with autoencoder leads to obtain optimized weights in the initial layer of CNN model. Moreover, features are enhanced in the HS images by utilizing the autoencoder. The CNN is used for efficient extraction of the features and same is also utilised for the classification of HS data. The potential of the proposed approach has been verified by conducting the experiments on three recent datasets. The experimental results are compared with the results obtained in the geoscience and remote sensing society (GRSS) Image Fusion Contest-2018 held at IEEE International Geoscience and Remote Sensing Symposium (IGARSS)-2018 and other existing frameworks for HSI classification. The testing accuracy of classification for the proposed approach is better than that of the other existing deep learning based methods.

Keywords: Autoencoder · CNN · Feature extraction · Hyperspectral classification

1 Introduction

Hyperspectral imagery (HSI) is a collection of images acquired in the spectral range of visible and near-infrared regions of the EM spectrum. It provides hundreds of spectral channels over a same captured geographical area [3]. Due to this, HS images are very useful in the classification and/or identification of the materials present in the captured scene. The HSI classification has been widely

© Springer Nature Singapore Pte Ltd. 2019
C. Arora and K. Mitra (Eds.): WCVA 2018, CCIS 1019, pp. 115–128, 2019.
https://doi.org/10.1007/978-981-15-1387-9_10

used in a variety of applications. However, it considered to be as a challenging task due to its nature of mixed pixels and large amount data volume [27]. This demands more efficient and robust techniques for HSI classification in order to extract features from the HSI data. In the last few decades, many researchers have attempted to address this problem. In the early phase, spectral domain classifiers such as multimodal logistic regression (MLR) [15], random forest (RF) [9] and support vector machines (SVMs) [22] were utilised for HSI classification. Recently, methods based on sparsity [5], Markov random fields (MRFs) [14] and morphological profiles (MPs) [2] are used for HSI classification and show the promising improvement over the traditional methods since they use both spatial and spectral details for HSI image classification.

In the field of computer vision, the deep learning based image classification has been an active area of research mainly due to the remarkable performance achieved through it. Specifically, in deep learning a lot of attention has been attracted by the convolutional neural network (CNN) which has an ability to extract features automatically from any kind of images. Also, it can be implemented as an end-to-end framework for multitask learning which is used in many applications such as multimedia search, vehicle detection, pedestrian detection and face detection. In addition to that, CNN can be widely used in the object localization [23], detection [24], recognition [13] and classification problem [11]. In comparison to the traditional image classification methods, the CNN can extract the features efficiently and classification map can be generated directly. In the literature, many methods have been proposed for classification of HSI data using different CNN models which aim to extract efficient high level deep features from HSI data [25]. Furthermore, multiple features can be learned simultaneously to extract more representative features. Learning of multiple features have been applied successfully in many image processing applications in computer vision field viz., multimedia search [17], vehicle detection [4], pedestrian detection [32] and face detection [8]. Furthermore, deep learning has become an area of interest of so many researchers and eventuated hot topic owing to simple strategy of algorithms, increase in speed and higher accuracy. In specific, a lot of attention has been attracted by convolutional neural network (CNN) due to its outstanding performance. CNN is used in so many domains such as object localization [23], object detection [24], solve the classification problem [11], object recognition [13], etc. In contrast to the traditional rule based methods for the feature extraction, CNN can extract deep features by learning hierarchical features from low level to high level. It has ability to extract features automatically from any category of images. Furthermore, CNN can be implemented as an end to end framework for multitask processing. Classification map can be directly generated using CNN. Therefore, various CNN models have been used for the classification of HSI.

Chen et al. [6] have introduced the first framework for HSI classification using many convolutional layers. This model extracts the invariant and nonlinear features from the hyperspectral images. Zhao et al. [30] have implemented a deep CNN model with dimension reduction algorithm for spatial-spectral feature extraction from hyperspectral image. In [20, 29], authors have proposed the HSI classification by constructing the CNN network with hierarchical feature extraction. Furthermore, Aptoula E. *et al.* [1] learn the several attribute profiles as an input to the CNN framework and stacked up on raw hyperspectral data. They efficiently classify the hyperspectral data by capturing its spectral and geometric properties. The aforementioned CNN models have focused on automatic spatial and spectral feature extraction. Zhao *et al.* [31] presented a different strategy that composite the extracted features by deep learning at multiple spatial scales which improves the performance of HSI classification at some extent. Selection of partial view strategy is utilized to combine the various view and applied as an input to the specific CNN architecture for land use classification [18]. Moreover, Xu *et al.* [26] presented a data classification of multi-source data. They use two tunnel CNN framework (i.e. extraction of spatial and spectral features separately by a single module) and fusion is performed on the multi-source data. Nevertheless, there are lack of investigations to classify various similar kind of features from HS image in the above work. Therefore, it is necessary to explore CNN framework which is able to extract spectral and spatial features simultaneously from various HS images and utilised the same for HSI classification.

In this paper, we use an autoencoder with CNN to improve the classification accuracy of the HSI classification. In the image classification, first it is necessary to enhance the features present from input hyperspectral image. Usually, an autoencoder can be used for feature enhancement due to its effective representation of the features at abstraction level with reduced dimensionality. Furthermore, the enhance feature map obtained through the autoencoder is fed to the CNN which extracts efficient features at each subsequent layer. The final classification is performed through the softmax classifier. Hence, the proposed framework includes simple architecture which does not require any post-processing module for classification. In proposed method, the CNN network consists two convolution layers with rectified linear unit (ReLU) activation function and three fully connected dense layers with leaky rectified linear unit (LReLU) and parametric rectified linear unit (PReLU). The final layer is implemented by including the number of classes of the HS image. The key contributions in this paper are as follow: (1) The proposed method consists shallow CNN network which is efficient and robust for HSI classification, (2) the proposed approach uses a CNN network with autoencoder in order to enhance the features of the HSI data and (3) the use of Adam optimizer in the proposed method makes the approach computationally efficient which is also well suited for large parameters. The rest of this paper is organized as follows: Sect. 2 introduces overall framework of the implemented method and also presents the detail discussion of the proposed model. The experimental results and discussions are presented in Sect. 3. Finally, Sect. 4 concludes the paper.

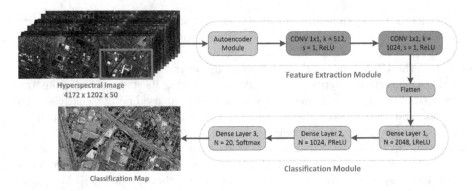

Fig. 1. Framework of the proposed method AECNN. Here, k, s and N indicate the number of filters, stride and nodes in the network, respectively.

2   Proposed Architecture Pipeline

In Fig. 1, we display the framework of the proposed method for classification of HS image. It consists of shallow architecture by using autoencoder and CNN. In the literature, it is proved that autoencoder is very efficient for the enhancement of the non-linear features on high dimensional data [7]. Moreover, pre-trained network give consistently better generalization. Our visualizations point to the observations that pre-trained networks learn qualitatively different features compared to networks without pre-training. The unsupervised pre-processing using autoencoder works as a regularizer that affects the starting point of the supervised training of the CNN model. Basically, unsupervised training using autoencoder favors the hidden units that compute features of the input x (input to CNN network) that correspond to major factors of variation in the true $P(x)$ i.e., probability of occurring of input x. Assuming that some of these are near features useful at predicting variations in y (actual output of CNN network), unsupervised pre-training sets up the parameters near a solution of low predictive generalization error. Additionally, it is also observed that with a small training set, usually researchers are not giving importance in minimizing the training error, because of overfitting issue. The training error is not a good way to distinguish between the generalization performances of two models. In that case, unsupervised pre-training helps to find apparent local minima that have better generalization error. Our main goal is to efficiently classify the densely connected classes of the HSI dataset (i.e. Houston). Dataset of Salinas and Pavia could perform better without use of an autoencoder but it could not be achieved by Houston dataset due to more number of classes with similar nature. Therefore, an autoencoder is used as a preprocessing task to obtain optimize weights at initial layer of CNN. Moreover, it is also observed that result of CNN module without using an autoencoder does not perform better as training attempts overfitting due to shallow architecture of CNN. As we increase the number of layers of CNN with various filter sets and number of channels to improve efficiency, it incredibly increases the complexity of architecture. The module without an autoencoder obtains less than 25% accuracy on Salinas and Pavia dataset, attains 17% accuracy on Houston dataset. Same module could obtain better accuracy by adding small preprocessing module (i.e. an autoencoder with three layers only). Approximately 100% reconstruction is obtained at the output of autoencoder and it is observed that the features are enhanced better than the original hyperspectral image. It acts as a kind of network pre-conditioner, putting the parameter values in the appropriate range for further supervised training. Therefore, accurate feature enhancement is done by pre-processing the hyperspectral image with the autoencoder. It also initializes the CNN model to a point in parameter space that somehow renders the optimization process more effective, in the sense of achieving a lower minimum of the empirical cost function. The replacement of autoencoder with CNN has also been verified experimentally and we found that CNN requires more layers with more number of kernels in order to obtain the same level of accuracy. Due to this, it also increases the complexity and training period of the architecture. Here, autoencoder takes an original data as input (x) and encode it in hidden layer (y) by mapping, $y = ReLU(Wx + b)$, where, W and b are the weight and bias of the autoencoder, respectively. The dimensionality of the intermediate latent representation is dynamic for all dataset due to large difference in size of them such as 2^{10}, 2^{15} and 2^{25} for Salinas, Pavia and Houston, respectively. Finally, it generates the output (x') by decoding the data from hidden layer (y) as, $x' = \sigma(W'y + b')$. Here, the reconstructed error is measured by using square error function as,

$$Loss(x, x') = ||x - x'||^2. \tag{1}$$

In the autoencoder, since the number of input nodes are same as the number of pixels in an HS image, the pixel's information is propagated through the network by discarding the nodes which are redundant. Finally, it reconstructs the original features at the output layer and obtains features with better enhancement. The output of the autoencoder i.e., x' is used as a input to the convolution layer of CNN.

The CNN consists two convolution layers with kernel of 1×1 to extract multiple features efficiently. First layer includes 512 kernels to extract deep features. A network without activation function works as a linear regression model which does not perform the given task in a better way. Hence, it requires to use non-linear activation function in order to learn complicated and complex form of data. Earlier, sigmoid function was used as an activation function but due to the vanishing gradient problem, now-a-days, rectified linear unit (ReLU) is used [21]. Hence, it is preferred to use at initial layer. The basic ReLU function is mathematically defined as, $f(x) = max(0, x)$. where, x represents the value of that particular node. ReLU describes that function is activated above the zero value; hence its partial derivative is one. Thus, the problem of vanishing gradient does not exist. Moreover, it saturates at zero which is more helpful to use as a input features. However, it has disadvantage during optimization due to zero gradient whenever unit is inactive. That means algorithm would never adjust the weights for initially inactive nodes. Again, number of kernels are increased by multiple of two of initial number of kernels to extract high level features at second layer i.e. 1024. Classification task demands more robust features to classify each object present in image. Hence, There are three dense layers are used. First, output data from convolutional layer is processed with 2048 nodes in the first dense layer. It could be possible that the learning becomes slow with training of ReLU function having zero gradient. This problem could be overcome by Leaky ReLu (LRelu) which is mathematically describes as,

$$f(x) = \begin{cases} x, & if \ x > 0; \\ 0.01x, & otherwise. \end{cases} \tag{2}$$

It allows small non-zero gradient when nodes are inactive and sacrifices the sparsity for the gradient during optimization which is more robust [19]. Moreover, parametric ReLU (PReLU) [10] makes the use of coefficient of leakage (a) into a parameter that is learned along with the other neural network parameters. It improves the model fitting with low computational cost and reduces the risk of over-fitting. Hence, second dense layer is implemented with 1024 nodes and data mapping is performed on that layer by utilizing the concept of PReLU which is described as,

$$f(x) = \begin{cases} x, & if \ x > 0; \\ ax, & otherwise. \end{cases} \tag{3}$$

Finally, softmax function is used at the output layer with the nodes having the total number of classes. The softmax layer is specifically used for classification problem to compute the probability of each classes. It can be expressed as,

$$P(\hat{y} = c|x) = \frac{exp(x^T W_c)}{\sum_{k=1}^{C} exp(x^T W_k)}. \tag{4}$$

Equation (4) describes the predicted probability of cth class, defined by given input vector x and weight vector W. The loss function in the network measures the performance of the classification. For more than two number of classes, categorical cross entropy is preferred and it can be calculated by separating the loss for each class label and then sum the result of individual as,

$$L(y, \hat{y}) = -\sum_{c=1}^{M} y_c log(\hat{y}_c). \tag{5}$$

Furthermore, the appropriate optimizer could enhance the results with better efficiency. Hence, Adam is preferred to optimize the model for the training of high dimensional hyperspectral image. Adam takes the benefits of adaptive gradient and RMS propagation by instead of adapting the learning rate parameter on the average, it uses average of the gradient's second moment [12]. Hence, it calculates an exponential moving average of the gradient and the squared gradient. It is calculated by,

$$W'_{ij+1} = W_{ij} - \frac{\eta}{\sqrt{\hat{v}_t} + \epsilon} \hat{m}_t, \tag{6}$$

where, $\hat{m}_t = \frac{m_t}{1-\alpha_1^t}$ and $\hat{v}_t = \frac{v_t}{1-\alpha_2^t}$. Where, $m_t = \alpha_1 m_{t-1} + (1 - \alpha_1)g_t$ represents the decaying average of past gradient and $v_t = \alpha_2 v_{t-1} + (1 - \alpha_2)g_t^2$ which is decaying average of past squared gradient, α_1 and α_2 are decay rate which is close to 1.

3   Experimental Results

The potential of the proposed method has been verified by conducting the experiments on three different HSI datasets: 1. Salinas, 2. Pavia and 3. Houston. We use Keras tensorflow libraries in order to implement the algorithm in a computer system configured with Intel (R) Core (TM) i7-7700 CPU @3.60GHz × 8, 32 GB RAM and a GPU NVIDIA GeForce GTX 1070 with 8-GB GDDR5. In the experiments, the pixel-wise annotation is performed using ERDAS Imagine and data is exported to generate .mat file for importing in the algorithm. The original dataset has been divided into two parts for training (70% samples) and testing (30% samples). All experiments have been conducted with batch size of 64. Epochs are set to 150 for Salinas and Pavia dataset, and 500 for Houston dataset. The classification performance of HSI data is measured in terms of accuracy and efficiency coefficient called kappa (κ) to validate the quantitative performance of the proposed method. Another performance evaluation parameters such as precision, recall rate and f1-score are also measured. To illustrate the performance of the proposed algorithm, we have also compared the quantitative performance of the proposed method with the different CNN based methods such as CNN-PPF [16], 3D-CNN [28] and the recently proposed two tunnel CNN (2T-CNN) method [26].

3.1   Experimental Evaluation on Salinas Dataset

First experiment is conducted on the Salinas dataset. It is acquired by the Airborne Visible Infrared Imaging Spectrometer (AVIIS) sensor which includes 224 spectral bands over Salinas, California. The size of the images in this dataset is 512×217 pixels with geometric resolution of 3.7 m. It contains 16 classes including vegetables, bare soils and vineyard fields. The classification map obtained using the proposed method for this dataset is displayed in Fig. 2 along with the original and its ground-truth images. The evaluation parameters such as precision, recall rate, f1-score and accuracy are measured for each classes. The overall classification accuracy and kappa coefficient are also evaluated and same are depicted in Table 1 for the proposed method along with the other existing HSI classification methods. We obtain 98.06 % accuracy and kappa of 0.9785 value using the proposed method which show the better performance of the proposed method over the recently proposed existing methods.

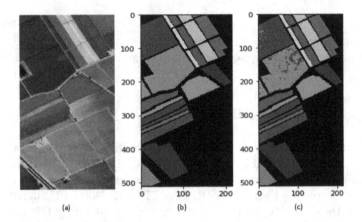

Fig. 2. The classification map obtained using the proposed method on Salinas dataset: (a) Salinas dataset, (b) ground-truth classification map and (c) classification map obtained using the proposed method.

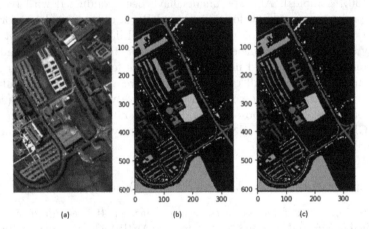

Fig. 3. The classification map obtained using the proposed method on Pavia University dataset: (a) Pavia University Dataset, (b) ground-truth classification map and (c) classification map obtained using the proposed method.

3.2 Experimental Evaluation on Pavia Dataset

Pavia University dataset is acquired by Reflective Optics System Imaging Spectrometer (ROSIS) sensor over university of Pavia, Northern Italy. It has 103 spectral bands ranging from 0.43 to 0.86 μ m with 610 × 340 pixels and the spatial resolution of 1.3 m. This dataset has 9 classes. Figure 3 shows the classification performance obtained using the proposed method on Pavia dataset. The accuracy and kappa measurement along with other evaluation parameters are displayed in Table 2 and they show that the proposed method obtains 99.95% accuracy with kappa value of 0.9993. Here, it is observed that the values of

Table 1. The different quantitative measures obtained using the proposed method for Salinas dataset.

Classes	Samples	Precision	Recall	f1-score	Accuracy(%)			
					Proposed	CNN-PPF [16]	3D-CNN [28]	2T-CNN [26]
Brocoli_green_weeds_1	209	1.00	1.00	1.00	**100**	100	98.36	100
Brocoli_green_weeds_2	372	1.00	1.00	1.00	99.05	**99.88**	98.43	99.09
Fallow	198	1.00	1.00	1.00	98.00	**99.60**	92.97	99.49
Fallow_rough_plow	117	1.00	1.00	1.00	**100**	99.49	99.46	99.16
Fallow_smooth	266	1.00	1.00	1.00	99.46	98.94	91.38	**99.68**
Stubble	361	1.00	1.00	1.00	**100**	99.97	99.83	100
Celery	370	1.00	1.00	1.00	**100**	100	99.68	99.64
Grapes_untrained	1131	0.99	0.90	0.94	**99.66**	88.68	68.94	91.36
Soil_vinyard_develop	643	1.00	1.00	1.00	**99.76**	98.33	98.45	99.45
Corn_senesced_green_weeds	330	1.00	1.00	1.00	**99.49**	98.60	73.31	97.66
Lettuce_romaine_4wk	107	1.00	1.00	1.00	**100**	99.54	90.85	100
Lettuce_romaine_5wk	180	0.99	1.00	1.00	**100**	100	98.31	100
Lettuce_romaine_6wk	106	1.00	1.00	1.00	**100**	99.44	97.43	100
Lettuce_romaine_7wk	90	1.00	1.00	1.00	**99.51**	98.96	94.76	99.00
Vinyard_untrained	739	0.87	0.98	0.92	**93.14**	83.53	63.75	88.00
Stone-Steel-Towers	191	1.00	1.00	1.00	99.73	99.31	89.83	**100**
Avg./Total	5413	0.98	0.98	0.98	**98.06**	94.80	85.24	97.72
κ					**0.9785**	0.9325	0.8360	0.9745

Table 2. The different quantitative measures obtained using the proposed method for Pavia dataset.

Classes	Samples	Precision	Recall	f1-score	Accuracy(%)			
					Proposed	CNN-PPF [16]	3D-CNN [28]	2T-CNN [26]
Asphalt	665	1.00	1.00	1.00	**100**	97.42	69.42	98.69
Meadows	1845	1.00	1.00	1.00	**100**	95.76	58.51	99.21
Gravel	215	1.00	1.00	1.00	**100**	94.05	78.86	99.05
Trees	309	1.00	1.00	1.00	**100**	97.52	99.02	99.05
Painted metal sheets	140	1.00	1.00	1.00	**100**	100	100	100
Bare Soil	488	1.00	1.00	1.00	99.79	99.13	63.35	**99.92**
Bitumen	146	1.00	1.00	1.00	**100**	96.19	93.82	99.93
Self-Blocking Bricks	385	1.00	1.00	1.00	**99.74**	93.62	57.54	97.99
Shadows	85	1.00	1.00	1.00	**100**	99.60	97.67	100
Avg./Total	4278	1.00	1.00	1.00	**99.95**	96.48	67.85	99.13
κ		-			**0.9993**	0.9582	0.6040	0.9883

these quantitative measures are better when compared to the same with other existing HS classification methods.

Table 3. The different quantitative measures obtained using the proposed method for Houston dataset.

Classes	Proposed method					2T-CNN [26]
	Samples	Precision	Recall	f1-score	Accuracy(%)	Accuracy(%)
Healthy grass	43072	0.92	0.96	0.94	**96.52**	83.38
Stressed grass	36804	0.92	0.92	0.92	**93.84**	84.21
Stadium seat	32564	0.75	0.72	0.74	**75.15**	–
Paved parking lot	71367	0.91	0.91	0.91	92.05	**92.51**
Unpaved parking lot	8953	0.83	0.76	0.79	78.69	**92.63**
Deciduous tree	49446	0.95	0.94	0.94	**94.28**	–
Bare earth	39114	0.83	0.84	0.84	85.93	**98.58**
Train	100206	0.84	0.90	0.87	**90.84**	–
Evergreen tree	102229	0.98	0.97	0.97	**97.21**	93.18
Artificial turf	44117	0.95	0.94	0.94	94.49	**99.60**
Road	13876	0.70	0.62	0.66	66.91	**78.66**
Highway	7465	0.74	0.61	0.67	**66.13**	52.90
Railway	70500	0.92	0.92	0.92	**93.15**	82.16
Car	16046	0.81	0.63	0.71	**66.87**	–
Residential building	17915	0.89	0.78	0.83	81.34	**85.45**
Non-residential building	32671	0.68	0.76	0.72	**79.46**	69.14
Major throughfare	9741	0.78	0.52	0.63	**56.17**	–
Sidewalk	53603	0.84	0.90	0.87	91.62	**99.79**
Unclassified	2523	0.59	0.46	0.52	**53.75**	–
Avg./Total	752212	0.88	0.88	0.88	**87.92**	84.08
κ					**0.8684**	0.8274

3.3   Experimental Evaluation on Houston Dataset

The third experiment is conducted on Houston dataset which is most recent HSI dataset with highest number of classes. The HS images in this dataset are acquired by National Center for Airborne Laser Mapping (NCALM) and they are made available through IEEE GRSS Data Fusion Contest 2018. It covers a 380–1050 nm spectral range with 50 bands at a 1-m ground sampling distance (GSD). It includes 20 classes with range from natural land cover (e.g., water, grass, tree and bare earth) to man-made objects (e.g., vehicles, roads and buildings). The classification results obtained using the proposed method for this dataset are depicted in Fig. 4. In this figure, boxes with different colors indicate that the selected samples are used for testing. The quantitative measurement parameters have been displayed in Table 3 and it shows that the proposed method obtains 87.92% testing accuracy and kappa of 0.8684 value which are highest among the other existing HSI classification methods. In addition to that, we have compared our results with winner of the GRSS Image Fusion Contest-2018

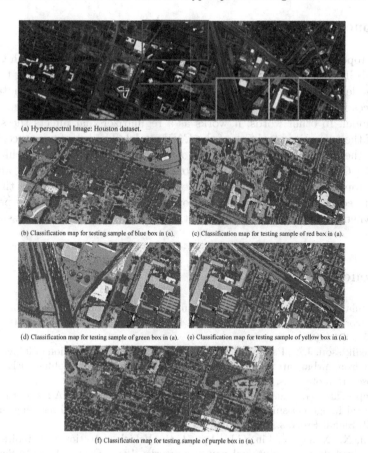

(a) Hyperspectral Image: Houston dataset.

(b) Classification map for testing sample of blue box in (a).   (c) Classification map for testing sample of red box in (a).

(d) Classification map for testing sample of green box in (a).   (e) Classification map for testing sample of yellow box in (a).

(f) Classification map for testing sample of purple box in (a).

Fig. 4. Classification map obtained using the proposed method on Houston dataset. Here, boxes with different colors indicate the testing samples and its classification maps are displayed in (b)–(f).

on HSI classification. They obtained 77.39% accuracy and kappa value of 0.7300 which proves that proposed method attains the better performance.

Since this dataset consists more number of classes, same is used to obtain the other quantitative measures such as precision, recall rate, f1-score, accuracy and number of detected samples of individual classes from total tested samples by using the proposed method which depicted in Table 3. The architecture in the proposed method is comparatively less complex than that of the other existing methods. Also, the hybrid concept of autoencoder and CNN provides better extraction of robust and deep features and same also helps to improve the efficiency of the proposed algorithm.

4   Conclusion

In the proposed method, we use a simple two layered autoencoder with CNN to classify HSI data. Due to the use of an autoencoder as a preprocessing task, the CNN model extracts the features efficiently from the hyperspectral data. This pre-processing procedure leads to obtain optimized weights in the initial layer of CNN model. In other words, it works as a regularizer that affects the starting point of the supervised training of the CNN model. The use of autoencoder also reduces the complexity as compared to the corresponding CNN architecture. Also, the CNN based feature extraction improves the classification accuracy of the proposed method. The experiments have been conducted on the three different recent HSI datasets which show that the proposed method performs better when compared to the same with the other existing HS image classification methods.

References

1. Aptoula, E., Ozdemir, M.C., Yanikoglu, B.: Deep learning with attribute profiles for hyperspectral image classification. IEEE Geosci. Remote Sens. Lett. **13**(12), 1970–1974 (2016)
2. Benediktsson, J.A., Palmason, J.A., Sveinsson, J.R.: Classification of hyperspectral data from urban areas based on extended morphological profiles. IEEE Trans. Geosci. Remote Sens. **43**(3), 480–491 (2005)
3. Camps-Valls, G., Tuia, D., Bruzzone, L., Benediktsson, J.A.: Advances in hyperspectral image classification: earth monitoring with statistical learning methods. IEEE Signal Process. Mag. **31**(1), 45–54 (2014)
4. Chen, X., Xiang, S., Liu, C.L., Pan, C.H.: Vehicle detection in satellite images by hybrid deep convolutional neural networks. IEEE Geosci. Remote Sens. Lett **11**(10), 1797–1801 (2014)
5. Chen, Y., Nasrabadi, N.M., Tran, T.D.: Hyperspectral image classification using dictionary-based sparse representation. IEEE Trans. Geosci. Remote Sens. **49**(10), 3973–3985 (2011)
6. Chen, Y., Jiang, H., Li, C., Jia, X., Ghamisi, P.: Deep feature extraction and classification of hyperspectral images based on convolutional neural networks. IEEE Trans. Geosci. Remote Sens. **54**(10), 6232–6251 (2016)
7. Chen, Y., Lin, Z., Zhao, X., Wang, G., Gu, Y.: Deep learning-based classification of hyperspectral data. IEEE J. Sel. Top. Appl. Earth Obs. Remote. Sens. **7**(6), 2094–2107 (2014)
8. Ding, C., Xu, C., Tao, D.: Multi-task pose-invariant face recognition. IEEE Trans. Image Process. **24**(3), 980–993 (2015)
9. Ham, J., Chen, Y., Crawford, M.M., Ghosh, J.: Investigation of the random forest framework for classification of hyperspectral data. IEEE Trans. Geosci. Remote Sens. **43**(3), 492–501 (2005)
10. He, K., Zhang, X., Ren, S., Sun, J.: Delving deep into rectifiers: surpassing human-level performance on imagenet classification. In: Proceedings of the IEEE International Conference on Computer Vision, pp. 1026–1034 (2015)

11. Karpathy, A., Toderici, G., Shetty, S., Leung, T., Sukthankar, R., Fei-Fei, L.: Large-scale video classification with convolutional neural networks. In: Proceedings of the IEEE Conference on Computer Vision and Pattern Recognition, pp. 1725–1732 (2014)

12. Kingma, D.P., Ba, J.L.: Adam: A method for stochastic optimization. In: Proceedings of 3rd International Conference for Learning Representations, pp. 1–15 (2015)

13. Lawrence, S., Giles, C.L., Tsoi, A.C., Back, A.D.: Face recognition: a convolutional neural-network approach. IEEE Trans. Neural Netw. 8(1), 98–113 (1997)

14. Li, J., Bioucas-Dias, J.M., Plaza, A.: Spectral-spatial hyperspectral image segmentation using subspace multinomial logistic regression and Markov random fields. IEEE Trans. Geosci. Remote Sens. 50(3), 809–823 (2012)

15. Li, J., Bioucas-Dias, J.M., Plaza, A.: Semisupervised hyperspectral image classification using soft sparse multinomial logistic regression. IEEE Geosci. Remote Sens. Lett. 10(2), 318–322 (2013)

16. Li, W., Wu, G., Zhang, F., Du, Q.: Hyperspectral image classification using deep pixel-pair features. IEEE Trans. Geosci. Remote Sens. 55(2), 844–853 (2017)

17. Liu, W., Mei, T., Zhang, Y., Che, C., Luo, J.: Multi-task deep visual-semantic embedding for video thumbnail selection. In: Proceedings of the IEEE Conference on Computer Vision and Pattern Recognition, pp. 3707–3715 (2015)

18. Luus, F.P., Salmon, B.P., Van den Bergh, F., Maharaj, B.T.J.: Multiview deep learning for land-use classification. IEEE Geosci. Remote Sens. Lett. 12(12), 2448–2452 (2015)

19. Maas, A.L., Hannun, A.Y., Ng, A.Y.: Rectifier nonlinearities improve neural network acoustic models. In: Proceedings of ICML, vol. 30, p. 3 (2013)

20. Makantasis, K., Karantzalos, K., Doulamis, A., Doulamis, N.: Deep supervised learning for hyperspectral data classification through convolutional neural networks. In: 2015 IEEE International Geoscience and Remote Sensing Symposium (IGARSS), pp. 4959–4962. IEEE (2015)

21. Nair, V., Hinton, G.E.: Rectified linear units improve restricted Boltzmann machines. In: Proceedings of the 27th International Conference on Machine Learning (ICML 2010), pp. 807–814 (2010)

22. Pal, M., Foody, G.M.: Feature selection for classification of hyperspectral data by SVM. IEEE Trans. Geosci. Remote Sens. 48(5), 2297–2307 (2010)

23. Parkhi, O.M., Vedaldi, A., Zisserman, A., et al.: Deep face recognition. In: BMVC, vol. 1, p. 6 (2015)

24. Ren, S., He, K., Girshick, R., Sun, J.: Faster R-CNN: towards real-time object detection with region proposal networks. IEEE Trans. Pattern Anal. Mach. Intell. 39(6), 1137–1149 (2017)

25. Windrim, L., Melkumyan, A., Murphy, R.J., Chlingaryan, A., Ramakrishnan, R.: Pretraining for hyperspectral convolutional neural network classification. IEEE Trans. Geosci. Remote Sens. 56, 2798–2810 (2018)

26. Xu, X., Li, W., Ran, Q., Du, Q., Gao, L., Zhang, B.: Multisource remote sensing data classification based on convolutional neural network. IEEE Trans. Geosci. Remote Sens. 56(2), 937–949 (2018)

27. Yan, D., Chu, Y., Li, L., Liu, D.: Hyperspectral remote sensing image classification with information discriminative extreme learning machine. Multimed. Tools Appl. 77(5), 5803–5818 (2018)

28. Yu, S., Jia, S., Xu, C.: Convolutional neural networks for hyperspectral image classification. Neurocomputing 219, 88–98 (2017)

29. Yue, J., Zhao, W., Mao, S., Liu, H.: Spectral-spatial classification of hyperspectral images using deep convolutional neural networks. Remote Sens. Lett. **6**(6), 468–477 (2015)

30. Zhao, W., Du, S.: Spectral-spatial feature extraction for hyperspectral image classification: a dimension reduction and deep learning approach. IEEE Trans. Geosci. Remote. Sens. **54**(8), 4544–4554 (2016)

31. Zhao, W., Guo, Z., Yue, J., Zhang, X., Luo, L.: On combining multiscale deep learning features for the classification of hyperspectral remote sensing imagery. Int. J. Remote Sens. **36**(13), 3368–3379 (2015)

32. Zhu, C., Peng, Y.: A boosted multi-task model for pedestrian detection with occlusion handling. IEEE Trans. Image Process. **24**(12), 5619–5629 (2015)

Author Index

Printed in the United States
By Bookmasters